Famous Planet Earth Caves

Vol. 2

Hermann's Cave (Germany) - A Late Pleistocene Cave Bear Den

Authored by

Cajus G. Diedrich
PaleoLogic, Research Institute,
Petra Bezruce 96, CZ-26751 Zdice
Czech Republic
www.paleologic.eu

General:

1. Any dispute or claim arising out of or in connection with this License Agreement or the Work (including non-contractual disputes or claims) will be governed by and construed in accordance with the laws of the U.A.E. as applied in the Emirate of Dubai. Each party agrees that the courts of the Emirate of Dubai shall have exclusive jurisdiction to settle any dispute or claim arising out of or in connection with this License Agreement or the Work (including non-contractual disputes or claims).
2. Your rights under this License Agreement will automatically terminate without notice and without the need for a court order if at any point you breach any terms of this License Agreement. In no event will any delay or failure by Bentham Science Publishers in enforcing your compliance with this License Agreement constitute a waiver of any of its rights.
3. You acknowledge that you have read this License Agreement, and agree to be bound by its terms and conditions. To the extent that any other terms and conditions presented on any website of Bentham Science Publishers conflict with, or are inconsistent with, the terms and conditions set out in this License Agreement, you acknowledge that the terms and conditions set out in this License Agreement shall prevail.

Bentham Science Publishers Ltd.
Executive Suite Y - 2
PO Box 7917, Saif Zone
Sharjah, U.A.E.
Email: subscriptions@benthamscience.org

BENTHAM SCIENCE

CONTENTS

FOREWORD

This book is focusing on a Late Pleistocene Weichselian Glacial cave bear den in north-central Europe. Such cave bear dens are larger cave systems, mostly filled up with ten thousands up to a half Million of cave bear bones. However, this book does not survey only the bones of the Hermann's Cave in the Harz Mountains, Saxony-Anhalt (northern Germany), but also highlights the paleoecological change of the den during 113.000 years located in the Bode River valley close to former valley glaciers present in the Last Ice Age. Interdisciplinary geological, paleontological and archaeological studies make the cave important for our knowledge on the local geomorphological development of the Harz Mountains during the last three Ice Ages (Elsterian, Saalian and Weichselian) reflecting the impact of the most northern mountain valley glaciers in Germany. Recent research during the past 10 years has changed dramatically our view on "the cave bear". New molecular data and osteometric evidence from skulls and teeth (= "cave bear clock") suggest that speciation processes resulted into the separation of several species and subspecies within the Late Pleistocene – the past 113.000 years. The behavior and ecological role of European cave bear taxa was consequently misunderstood for a long time and the antagonistic interactions with the other extinct top predators - steppe lions, Ice Age leopards, Ice Age spotted hyenas, and Ice Age wolves – have not been fully considered. The predation and scavenging of carcasses representing a smaller and a larger cave bear species/subspecies are explained in this book as spectacular examples of the interaction with steppe lion whose skeletons (at least three) were discovered deeper in the Hermann's Cave. This extraordinary book on Hermann's Cave provides clear evidence for its international paleoecological importance. It deepens our knowledge and understanding of life and extinction of cave bear taxa which appears as a series of events beginning with changing climate leading to the expansion of boreal forests and the reduction of food resources during in the Last Glacial Maximum, and followed by the predatory impact of large carnivores and Cromagnon humans. In the deeper part of the cave Aurignacien hunters left with their tools the oldest cave bear hunt proves most probably with propulsion weapon technique during the bears′ winterly inactivity period.

Prof. Dr. U. Sinsch
University of Koblenz-Landau, IfIN, Deptt. of Biology,
AG Zoology, Universitätsstr. 1
D-56070 Koblenz,
Germany
Email: sinsch@uni-koblenz.de

PREFACE

The new Famous Planet Earth Caves e-book series challenges to present important caves all over the world scientifically, but somehow also popular in a mixture, that non-experts can understand their main importance. The Hermann's Cave in northern Germany is one of the largest show caves in Germany and Europe and has an important Ice Age related history starting only about 350.000 years ago with most importance for the Late Pleistocene (MIS 5d-2) and its impact for the landscape and LGM glacier reconstructions of the Harz Mountain Range. This is a beautiful granite (Brocken Peak) and limestone (cave areas) rock and valley cutted landscape, that became famous already in historic times. It was visited by many famous natural scientist pioneers starting already at the end of the 17th century. The book is interdisciplinary but with a strong focus on its main international importance – the Late Pleistocene fauna and archeology. The cave system was used by different cave bear types as cave bear den within a former boreal forest environment of the medium high up to 1.014 a.s.l. elevated most northern German mountain range. It presents a review of scientific work of the Brunswick Palaeontologist/Archaeologicst pioneer Prof. Dr. W. Blasius after more than 200 years. It is a catalogue compendium and adds knowledge to small and large cave bear species/subspecies ecology in Europe including interactions with steppe lions and Cromagnon humans, and their life and battle to survive in taiga forest mountain areas of central Europe. In Europe it is the second very important cave, that has delivered some steppe lion skeletons deeper in the cave between cave bear skeletons – a rare proof for their cave bear predation. From the archaeological point of view, the cave has another unique and convincing European cave bear hunting prove by Late Palaeolithics deeper in the cave using propulsion weapons. The project also demonstrates the large "cave society problem" and the conflicts between scientists, private caver clubs and cave owners. Spelunkers have other interests as bat protectors, palaeontologists, archaeologists or cave owners, whereas the bones which are the most important in the cave – are treated worst: being trampled, skeletons are damaged in context by taking trophies or during electric cable installations, *etc*. The best/worst example for the long-tradition of quite bad cave treatment is the original cave bear skeleton composite made historically by Kloos, which was replaced "dumped" in the Bear Hall between other bones by a "plastic cave bear skeleton". The "palaeontological Picasso" was of less worth, at a cost of 5.000 Euro most expensive replica, and it was worth less as the 2015 expensively restaurated "concrete bear" in the centre of Rübeland village – which demonstrates the still remaining difficulties in understanding the real value of the cave and its "stories". This book should open the eyes for the really important facts of the cave and its "bone stories", so that they might become "alive" in an Ice Age cave Safari story in a future Museum being installed sensitively within the cave. Finally, the book is dedicated to a good friend, the archaeological cave researcher and specialist for Iron Age cave use in the Sauerland Karst cave region of northern Germany, Dr. Wilhelm Bleicher, who sadly passed away beginning of 2016.

Cajus G. Diedrich
PaleoLogic, Research Institute,
Petra Bezruce 96, CZ-26751 Zdice,
Czech Republic
Email: cdiedri@gmx.net

ACKNOWLEDGEMENTS

This book project was supported by the new cave manager T. Schult, which is the base for a future new exhibition and "cave story". The city Oberharz am Brocken (here especially mayor F. Damsch) and the Landkreis Harz, Untere Naturschutzbehörde (permit Nr. 96705-2012/kkr) such as former cave manager C. Hensel allowed the cave studies. Prof. Dr. U. Joger gave access to the historical material from the Baumann's Cave which was supported by the curator Dr. R. Kosma of the Staatliches Naturhistorisches Museum Braunschweig. Access to the collection of the Museum für Naturkunde Berlin of Humboldt University with the Hermann Cave material (Steiner 1961/Völker 1984/85 excavations) was possible thanks to PD Dr. O. Hampe and Mrs. S. Scheffel. A bone cleaning of about 2.500 bones and cave expedition was assisted by N./E. Schuhose, G. Vollmer and B. Baumbach of the Arbeitsgemeinschaft für Karstkunde Harz e.V. Dr. R. Mischka (Landesamt für Denkmalpflege und Archäologie Sachsen-Anhalt) supported the study of the archaeological and palaeontological finds which are housed in the LDA. Dr. B. Stoll-Tucker supported with a photo of the Late Palaeolithic bone point. Finally, I thank Prof. Dr. B. Rothschild review support of one chapter. A good friend, the biologist T. Parker from the US, made a great spell-check of the first version.

The digital 3D-illustrations were made thankfully by the palaeoartist Roman Uchytel (www.prehistoric-fauna.com).

CONFLICT OF INTEREST

There is a competing interest with the show cave owners (city Oberharz), which will need to use the content of the book for commercially show cave guided tours and exhibitions. All contents or illustrations stay in their commercial copyrights by the author and have to be requested for use permit.

COLLECTIONS

The material was spread over several institutions: Landesmusuem Denkmalpflege und Archäologie Halle/Saale (= **LDA**: excavations by Steiner), Landesmuseum für Naturkunde Braunschweig (= **LMNB**: excavations by Kloos and Blasius), Naturkundemuseum der Humboldt-Universität Berlin (= **MB**: excavation by Völker), Museum Bernburg (= **MBE**), Museum Quedlinburg (= **MQ**), and Rübeland Caves (= **RC**) - there, partly in-situ.

CHAPTER 1

Introduction

Abstract: The northern German Hermann's Cave below the 1.141 a.s.l meters high Brocken peak in the centre of the Harz Mountains (Saxony-Anhalt) was discovered by chance during road construction work at the village Rübeland in the year 1866. It was explored and a first map was presented with first cave descriptions 1889 by the geologists Prof. Dr. J.H. Kloos and Prof. Dr. M. Müller from the Brunswick University. In 1890 the cave became one of the historically opened European show caves and is one of the largest tourist caves in Germany counting about 75.000 visitors per year. Palaeontological and archaeological pioneering research was made by the biologist Prof. Dr. W. Blasius from the Brunswick Natural History Museum, who was active from 1892 to 1901 with opening a small museum in front of the cave. The Hermann's Cave bear den belongs to one of the three bone-rich important and most northern European cave bear den sites being situated opposite the Baumann's Cave and not far from the Unicorn Cave in the Harz Mountain. In contrast to the other two mentioned caves, which were used by cave bears and carnivores such as Neanderthals in the Middle Pleistocene, the Hermann's Cave was accessible for cave bears, carnivores and even humans due to the starting pre-LGM glaciations of the Brocken peak and cave entrance collapse processes only in the Late Pleistocene. At this time, Ice Age steppe lions and Late Palaeolithic Cromagnon humans hunted different cave bear species deep in the cave.

Keywords: Hermann's Cave, Harz Mountain, northern Germany, large show cave, discover and excavation history, Kloos and Blasius, cave bear den, Late Pleistocene, lions and Late Palaeolithic hunters.

CAVE BEAR DENS IN GERMANY

The cave bear den overview map for the 37 larger and important recently known German Late Pleistocene cave sites (Fig. **1A**) [1] shows clusters in the cave-rich limestone/dolomite karst areas, where their bones can be preserved well in cave sediments.

The northwestern German Sauerland Mountains (North Rhine-Westphalia) Palaeozoic (Devonian) limestone karst caves (Fig. **1A**) have some important larger Late Pleistocene cave bear den sites with the Perick Caves, followed by the Keppler Cave [2], which latter was sadly already historically destroyed. There are several smaller cave bear den sites which delivered in historic times some to

Cajus G. Diedrich

several hundreds of bones such as the Bilstein Cave, Pfefferburg Cave or the Rösenbeck Cave [2]. One of the very small cave bear den sites of importance is the Teufelskammer Cave besides the famous Neanderthal Cave in the Neanderthal Valley [3]. Only the Middle to Late Palaeolithic Neanderthal and Cromagnon human camp Balve Cave overlaps with a cave bear den [4].

A smaller karstic region in Thuringia has mostly several small caves and cavities (Fig. **1A**). The largest cave and cave bear den there is the Altenstein Cave [5]. A smaller cave bear den cave with the Ilsen Cave [6] and very small den caves often overlap with Neanderthal-modern human Palaeolithic camps and hyena or wolf dens are in the Palaeozoic (Permian) dolomite area middle German Thuringian Mountains (Thuringia), such as the Wüste Scheuer Cave [6].

Most cave-rich is the Mesozoic dolomite (Late Jurassic) karstic Upper Franconia Mountain region of northern Bavaria. There the most famous "bone cave of Europe", the Zoolithen Cave is situated nearby Burggeilenreuth, which was started to be excavated already before 1800 [7]. Museums material comprises from this cave several Pleistocene holotype skulls of Late Pleistocene "cave bear", lion, hyena or the red wolf [7]. Further cave bear dens of larger size and with Late Pleistocene cave bear remains are the Grosse Teufels Cave, or the intensive studied Sophie's Cave, which was presented as first larger description of a German cave bear den cave in **Vol. 1** of this book series [1].

In southwestern Germany Baden Württemberg the karstic Mesozoic limestone (Late Jurassic) cave bear dens overlap in many cases with Palaeolithic camp sites [8]. Important cave bear dens are the Bärenhöhle Cave, the Sybillen Cave, or the Petersfels Cave [9].

HARZ MOUNTAIN CAVE BEAR DEN CAVES

Another cave-rich karst region is the Palaeozoic (Devonian) limestone area of the central northern German Harz Mountains in Saxony-Anhalt/Lower Saxony which has three Middle to Late Pleistocene cave bear dens [10 - 12] (Figs. **1B-C**). The map shows all three caves and to those important and fauna related open air sites surrounding the Harz Mountains such as the Late Pleistocene Westeregeln hyena den/Neanderthal camp [13], Late Pleistocene Seeweckenberge hyena den [14], herein added Late Pleistocene hyena den open air site Bennungen (Fig. **1C**), or close to the latter the much older early Middle Pleistocene Edersleben steppe mammoth open air site [15] in a time frame between 800.000 – 24.000 years ago during the Marine Isotope Stages 21 to 3. In several cases, there is an overlap of large carnivore den or human camp sites (Fig. **1C**) [10]. At some of the open air sites such as Bennungen ort Westeregeln (Fig. **1C**), at each only a few cave bear tooth remains have been found, which are important in the understanding of the

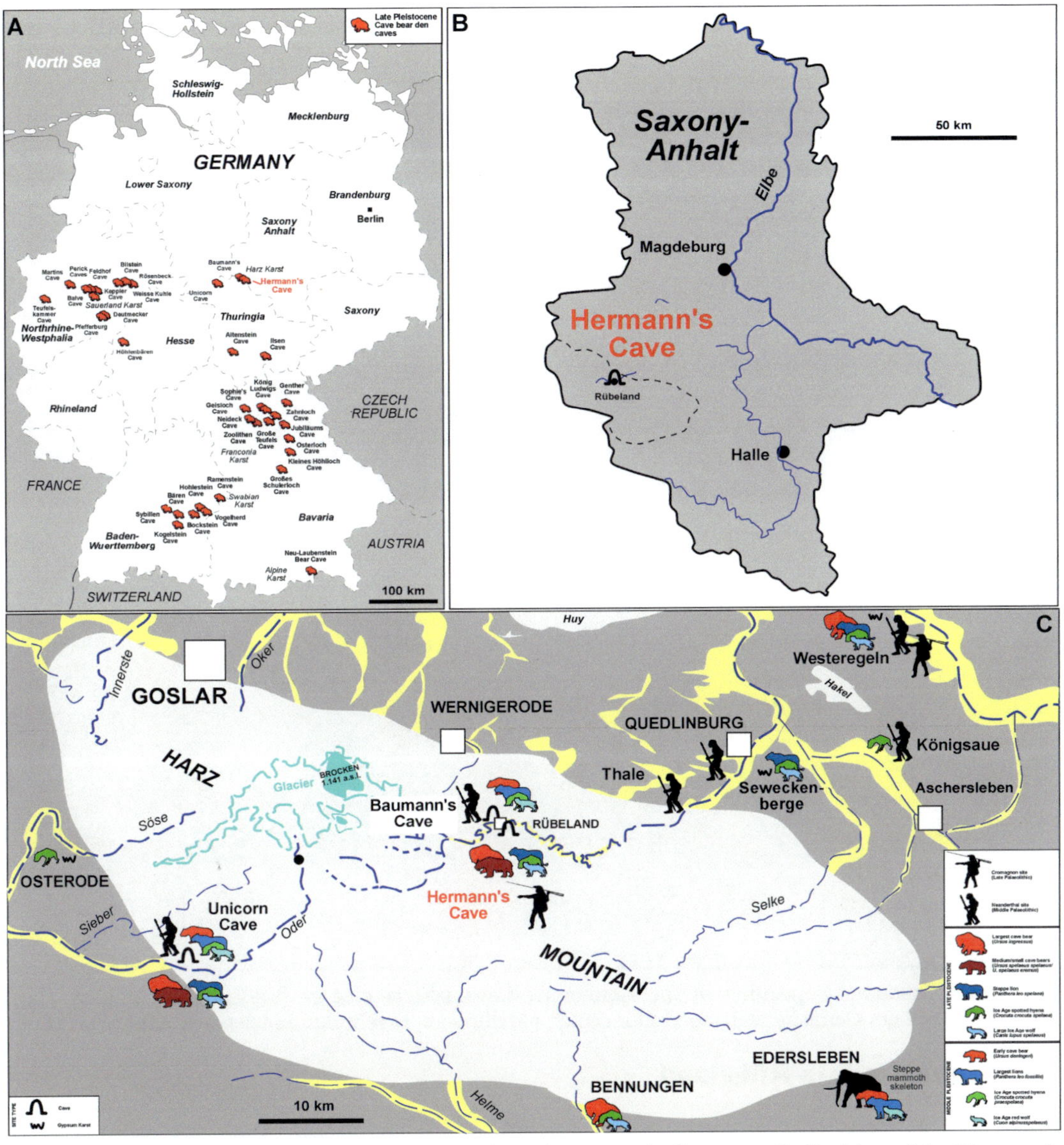

Figure 1. A. Late Pleistocene (MIS 3-5e) cave bear den caves in Germany. B. Position of the Hermann's Cave in Saxony-Anhalt. C. Important Middle to Late Pleistocene cave and open air megafauna or human camp sites in and around the Harz Mountain (composed from [10 - 12] and updated herein).

palaeogeographic distribution, seasonal migrations from the mountains to the lowlands or possible hunt by large predators, Neanderthal or modern humans. The older cave bear dens with rich bone material are the Unicorn Cave which is situated on the southern slope of the mountain chain (Fig. **1C**) and the Baumann's Cave. Both were used already in the Cromerian to Saalian Glacial and Interglacial periods as hibernation caves by older cave bear ancestors such as the small *Ursus deningeri* [10 - 12].

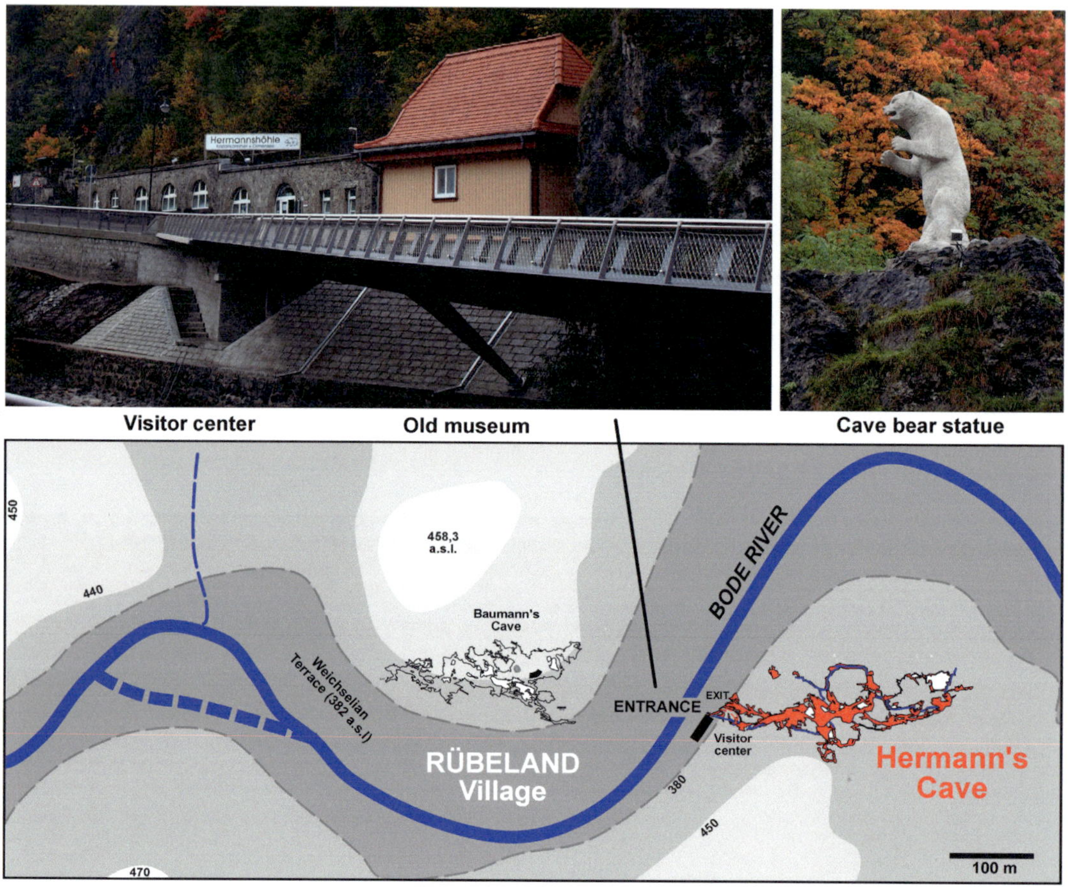

Figure 2. Topographic position of the Hermann's Cave system east of the Rübeland village, Harz Mountains, northern Germany with the Visitor center, old museum, new bridge and concrete cave bear statue.

Hermann's Cave in Rübeland

Opposite to the Baumann's Cave in the eastern valley part of the village Rübeland, there is the Hermann's Cave as the second and later discovered show cave (Figs. **1C, 2**). It had an average of 73.600 visitors per year between 2011-2015 [16]. The Google Earth coordinates are 51° 45′ 17″ N, 10° 50′ 50″ E whereas the today's entrance is at the elevation of about 394 meters a.s.l. It has several ponor elevations within 39 meters of elevation distances and is reaching 90 meters deep with its visitor loop into the Blue Chamber (Fig. **3**) [17]. To **Vol. 1** of this book series, this northern German and most northern European large cave bear den can be added as international important Ice Age fossil and archaeological site, which completes not only the knowledge about Late Pleistocene extinct cave bears of Europe. It furthermore improves the knowledge of Late Palaeolithic Aurignacien cave bear hunters and Late Pleistocene steppe lion hunting behavior.

Discovery and Research History

Only a few important historical data are repeated from the original historical sources, especially those concerning excavations and research published by Grotrian [18], Kloos and Müller [19] and Blasius [20 - 23] and from younger times. Their detailed reports are important to reconstruct and understand the places of excavated material, especially the largest Kloos/Blasius collections which are kept today in the Naturhistorische Museum Braunschweig [24]. There were further non-documented excavations or removals of skulls and larger bones within the long time-frame also in Socialistic times, and during times of poor cave management, which have no documentation. Some material appeared unexpected during the studies in collections of the Museum Bernburg or University Halle/Saale, other skulls and "trophies" seem to have been lost. Even modern, cave explorations still destroy parts of the bone layers, whereas Palaeontologist are not included for palaeontological monument work.

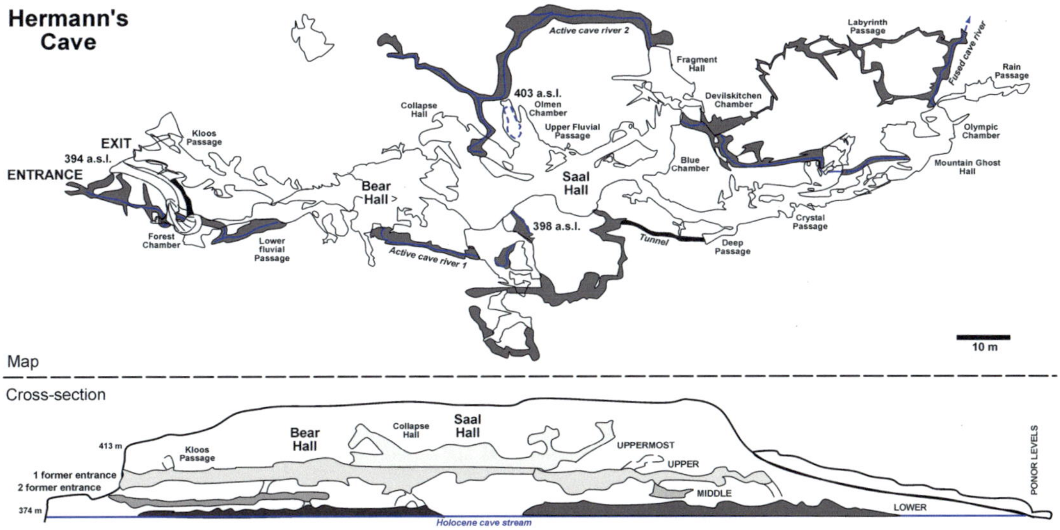

Figure 3. Map of the Hermann's Cave system with its main passages, halls and chambers and four glacial Bode River Valley connected ponor levels.

In 1866, the cave was discovered during construction work along the road from Rübeland to Hasselfelde, on the 28th of June by the road worker Wilhelm Angerstein, which was also named Sechserding. The cave was first called "Sechserdinghöhle" [18].

In 1875/76, first larger excavations were made by the Kammerrat Hermann Grotrian who renamed the cave to its today's name, Hermannshöhle (= Hermanns Cave) [18].

In 1889, the geologists Professor Dr. J.H. Kloos and Professor Dr. M. Müller presented the "monograph" of the Hermanns's Cave (Fig. **3**) including a first cave map [19], but excluding any illustrations of collected material.

In the year 1892 a small cave museum was planned. The Museumsinspektor Fritz Grabowski and the cave guide Fritz Hase made further excavations at several places in the cave. The largest and deepest excavation was made in the Bear Hall. There, the first Palaeolithic blade fragment was discovered, as well as a few woolly rhinoceros or reindeer bones, and a red deer antler tip. The extraordinary large amount of cave bear bones was also mentioned. Some of the material was exposed in the same year in the small Rübeland Museum, which is opposite to the today's entrance and visitor center [20].

Six years later in 1898 the biologist Professor Dr. Wilhelm Blasius (Fig. **3**) described the first remains of the presence of the Palaeolithic human in the Rübeland caves with securing evidence of a flintstone blade [21].

Excavations in 1901 by Blasius in the anterior part of the cave, about 8 meters distance from the former still blocked entrance [22]. He targeted in on the area of the first chamber (here named as Kloos Chamber which is besides the today's exit tunnel, Fig. **4**) with the discovery of small bones, especially birds and other remains, which are from the younger periods after the cave bears had already disappeared.

In 1961, the archaeologist Steiner [25] opened three smaller excavation areas. Two trenches were made at the southern side entrance to the Bear Hall, a third surface excavation of few square meters in the Sall Hall. None of those 76 lion bones described herein for the first time were identified as such which have most recently been found by the author in-between the "cave bear bone" collection.

1969 marks a period where the palaeontologist Dr. Schütt tried to describe and revision of the "Rübeland" caves fauna, but was not able to separate the Hermann's Cave from the Baumann's Cave fauna [26], in which several determinations were incorrect [24].

During the mid-eighties (1984/85) the geologist Dipl.-Geol. R. Völker made the few square meters wide excavation besides the ones of Steiner in the Sall Hall [27]. He reached about two meters deep into the pre-Late Pleistocene ponor river sands and gravels.

In 1994/95, the cave bear remains from the Völker excavation were published by Fischer as "cave bear *Ursus spelaeus*" whereas *e.g.* a patella of a lion (= spiny form) was included in those [28] some others were not identified as such (4 ribs, sterna bone, patella see Chapter lions). Other lion remains which are all from the small excavation area of Völker, were preliminary published [29], whereas the skull was overlooked as such (figurations only of maxillary pieces) and its parts were not fixed to a single skull.

In 2004, some of the stone tools from the Steiner excavation were identified by the archaeologist PD Dr. T. Weber to originate from Aurignaciens [30].

In 2005, the geologist Dr. W. Rosendahl took cave bear tooth and bone "samples" from the "bone dump" of the Hermann's Cave which were used to "date the cave bears of the Rübeland Caves" [31]. Also Nordhoff took speleothem samples without any stratigraphic or other cave genesis context [32], whereas some of his very useful data were integrated in the newer research [10] and published herein.

In 2006, the Late Palaeolithic bone point was dated with destroying a 5 mm large round part in its center by the initiative of the archaeologist Dr. J. Grünberg resulting in an AMS absolute age of 29.210 ± 210 BP [33].

Wilhelm Blasius (1845-1912)

Figure 4. Hermann's Cave palaeontological pioneer Professor Dr. W. Wilhelm Blasius from the former Naturhistorische Museum Braunschweig and one of his original written bone label notices from 1889 (coll. archive LMNB).

New Cave Research

In 2012, the Author explored the complete cave, documented sections, mapped some bonebed surfaces, and recovered all "bone dumps", especially the one in the Bear Hall. Within rescuing, also some of the dumped material in the Sall Hall (?Steiner/Völker excavations) important discoveries were bones belonging to steppe lion. Of this discovery 1.430 osteologically by bone types and species determinable and about 1.000 bone fragments such as about 20 skull (partly from Kloos excavations) were washed (many had algae growth on the surfaces), dried and coarsely inventoried such as stored in boxes for future exhibitions. This material, 2.280 bones and some skulls (including a lion skull) from the Völker excavation, 1.323 bones (including 8 skulls and many overlooked postcranial lion bones) build the main base for this book presentation.

In 2013, the first results were partly published [10, 24]. Herein, for the first time, all lion material is presented monographically, such as a review of all published and unpublished archaeological remains, whereas the cave bears will be discussed in more detail.

REFERENCES

[1] Diedrich C. Sophie's Cave - a Late Pleistocene cave bear den. Bentham 2015; p. 192.

[2] Diedrich C. Die letzten Hyänen in Westfalen - Neandertaler-Eiszeitsafari in Bottrop. Hohenlimb Heimatbl 2015; 1: 31-4.

[3] Diedrich C. The *Crocuta crocuta spelaea* (Goldfuss 1823) population and its prey from the Late Pleistocene Teufelskammer Cave hyena den besides the famous Palaeolithic Neanderthal Cave (NRW, NW Germany). Hist Biol 2011; 23(2/3): 237-70. [http://dx.doi.org/10.1080/08912963.2010.530348]

[4] Diedrich C. Periodical use of the Balve Cave (NW Germany) as a Late Pleistocene *Crocuta crocuta spelaea* (Goldfuss 1823) den: Hyena occupations and bone accumulations *vs.* human Middle Palaeolithic activity. Quat Int 2011; 233: 171-84. [http://dx.doi.org/10.1016/j.quaint.2010.02.027]

[5] Braniek G. Die Funde von Resten des Höhlenbären (*Ursus spelaeus* Rosenmüller 1794) aus den Höhlen von Altenstein und Bad Liebenstein. Schr Thüringer Landesanst Umw Geol 2002; 55: 101-34.

[6] Diedrich C. Late Pleistocene spotted hyena den sites and specialized rhinoceros scavengers in the Thuringian Mountain Zechstein karst (Central Germany). Quat Sci J 2015; 64(1): 29-45.

[7] Diedrich C. Pleistocene Stratigraphy and bone taphonomy in the Zoolithen Cave - and new theory about Esper's "biblic flood" with a first geomorphological Wiesent Valley and river terrace evolution model -? small glaciers and high/postglacial flood events in Upper Franconia. Eisz Gegenw. Quat Sci J 2013; 62(2): 162-74.

[8] Münzel SC, Stiller M, Hofreiter M, Mittnik A, Conard NJ, Bocherens H. Pleistocene bears in the Swabian Jura (Germany): Genetic replacement, ecological displacement, extinctions and survival. Quat Int 2011; 245: 225-37. [http://dx.doi.org/10.1016/j.quaint.2011.03.060]

[9] Kaulich B, Rosendahl W. The Neonate cave bear skeleton from the Petershöhle near Velden (Franconian Alb, Germany). Abh Karst- Höhlenk 2002; 34: 12-6.

[10] Diedrich C. Impact of the German Harz Mountain Weichselian ice-shield and valley glacier development onto Palaeolithics and megafauna disappearance. Quat Sci Rev 2013; 82: 167-98. [http://dx.doi.org/10.1016/j.quascirev.2013.08.013]

[11] Schütt G. Die cromerzeitlichen Bären aus der Einhornhöhle bei Scharzfeld. Mitt Gol Inst Techn Hochsch Hannover 1968; 7: 1-21.

[12] Nielbock R. Faunen des Eiszeitalters - Funde und Grabungen in Schlotten und Höhlen des Südharzes. Ber Niedersächsische Natursch-Akad 1998; 98(2): 61-70.

[13] Diedrich C. Late Pleistocene *Crocuta crocuta spelaea* (Goldfuss 1823) clans as prezewalski horse hunters and woolly rhinoceros scavengers at the open air commuting den and contemporary Neanderthal camp site Westeregeln (central Germany). J Archaeol Sci 2012; 39(6): 1749-67. [http://dx.doi.org/10.1016/j.jas.2012.01.013]

[14] Diedrich C. Seltene Freilandfunde der späteiszeitlichen Fleckenhyäne *Crocuta crocuta spelaea* (Goldfuss 1823) in Sachsen-Anhalt und Beitrag zu Horsttypen der letzten Hyänen im Jung-Pleistozän von Mitteldeutschland. Z Mitteldt Vorgesch 2014; 94: 1-19.

[15] Garutt VE, Nikolskaya VN. Über das Skelett vom Steppenelefant aus Edersleben. Beiträge zur Heimatforschung. Spengler Museum Sangerhausen 1988; 9: 3-13.

[16] https://de.wikipedia.org/wiki/Hermannsh%C3%B6hle_(Harz)

[17] Fricke U, Knolle F. Naturpark Harz - die Rübeländer Tropfsteinhöhlen. Goslar: V. Schadach 1999; p. 80.

[18] Grotrian H. Über das Vorkommen von Knochenresten der quartären Säugethier-Fauna in den Höhlen und Spalten des Devonkalkes von Rübeland am Harz. Z dt Geol Ges 1878; 30: 552-3.

[19] Kloos JH, Müller M. Die Hermannshöhle bei Rübeland. K. Schwier-Verlag, Weimar, 1889; p. 76.

[20] Blasius W. Neue Knochenfunde in den Höhlen bei Rübeland. Sitzungsb Ver Naturwissensch Braunschweig 1890; pp. 1-8.

[21] Blasius W. Spuren Paläolithischer Menschen in den Diluvial-Ablagerungen der Rübeländer-Höhlen Festschrift zur 29. Vers dt Anthropol Ges Braunschweig 1898; 1-37.

[22] Blasius W. Bericht über die fortgesetzten Ausgrabungen in den neuen Theilen der Baumannshöhle bei Rübeland im Harz während der Jahre 1898 und 1899. Jb Ver Naturwissensch Braunschweig 1901; 12: 83-6.

[23] Blasius W. Die Vogelfaunen in den diluvialen Ablagerungen der Rübeländer Höhlen. J Ornithol 1901; 122: 57-60. [http://dx.doi.org/10.1007/BF02206272]

[24] Diedrich C. Evolution, Horste, Taphonomie und Prädatoren der Rübeländer Höhlenbären, Harz (Norddeutschland). Mitt Karst- Höhlenf 2013; 59(1): 4-29.

[25] Steiner U. Zwei neue paläolithische Artefakte aus der Hermannshöhle bei Rübeland /Harz. Ausgrab Funde 1964; 9: 17-8.

[26] Schütt G. Die jungpleistozäne Fauna der Höhlen bei Rübeland im Harz. Quartär 1969; 20: 79-125.

[27] Völker R. Die Grabungen auf Bärenknochen im Frühjahr 1984 in der Hermannshöhle Rübeland. Ber Mitt Zentr Fach Höhlen- Karstf 1985; 85(1): 5-7.

[28] Fischer K. Der Höhlenbär *Ursus spelaeus* Rosenmüller and Heinroth 1794 von Rübeland. Braunschweiger naturkundliche Schr 1995; 4(4): 709-42.

[29] Fischer K. Neufunde von jungpleistozänen Höhlenlöwen *Panthera leo spelaea* (Goldfuss 1810) in Rübeland (Harz). Braunschw Naturkdl Schr 1994; 4(3): 455-71.

[30] Döhle HJ, Weber T. Höhlenbewohner in Mitteldeutschland? Die Funde aus Rübeland. Katalog zur Dauerausstellung im Landesmuseum für Vor- und Frühgeschichte Halle 2004; 1(1): 197-200.

[31] Rosendahl W, Döppes D, Joger U, *et al.* New radiometric datings of different cave bear sites in Germany - results and interpretations. Bull Soc Hist Nat Toulouse 2005; 141(1): 39-46.

[32] Nordhoff P. Stable isotope investigations on speleothems from different cave systems in Germany Unpublished Dissertation thesis, Georg-August-University Göttingen, 2005; p. 149.

[33] Grünberg J. New AMS dates for Palaeolithic and Mesolithic camp sites and single finds in Saxony-Anhalt and Thuringia (Germany). Proc Prehist Soc 2006; 72: 95-112. [http://dx.doi.org/10.1017/S0079497X00000797]

CHAPTER 2

Geology and Late Pleistocene Glacial Landscape

Abstract: The Hermann's Cave was formed into Late Devonian reef limestones of the geological Elbingeröder Complex starting during the Early Pleistocene in the central part of the Harz Mountain Range. Older river terrace sand/gravel deposits within the Hermann's Cave were deposited in the 16-20 meters higher elevated Saalian formed middle ponor level. Those are overlain by autochthonous younger early-middle Late Pleistocene (MIS5d-3) cave clays which contain the cave bear and other fauna such as archaeological remains. The sediments are dated indirectly by the "cave bear clock" (cave bear species, tooth and skull morphotypes) and by the absolute dated covering speleothem layer that cemented the uppermost cave bear skeletons and bones. This youngest speleothem phase produced candle stalagmites all over the cave system and levels before the LGM (MIS 2), and especially at the final Late Pleistocene (Bölling/Alleröd/Dryas; MIS 2/1 boundary) cold humid phases. In the warmer humid interstadial times of the early Late Pleistocene (MIS 5a and c), the today larger speleothems started to built up, which means that those are not older as 104.000 years.

Keywords: Devonian limestone, Karst, Hermann's Cave, Ponor stages, Three main levels, Middle to Late Pleistocene, Sediment types, Speleothem ages and forms.

HARZ LIMESTONE KARST GEOLOGY

The smaller of two Devonian limestone karst areas in the Harz Mountain (Fig. **1A**) is in the Iberg region, the second larger one with the Hermann's Cave is found the so-called Elbingeröder Complex [1 - 4] (Fig. **1B**). This geological zone is extended about 18 to 4.5 km and characterized by the early Middle Devonian siliciclastic marine shist (Wissenbach shist) and lower, middle and upper volcanic rock series [1 - 4] (Fig. **1B**). Those submarine volcanites are schalstein, spilits, keratophyrs and diabases, such as iron ore. Those built a rock series between 500-1.000 meters in thickness [1 - 4] (Fig. **1B**). During less volcanic time at the Middle Devonian Eifel/Givet boundary, massive stromatopore reefs and lime mud started to deposit up in shallow warm marine palaeoenvironments [1 - 5] (Fig. **1B**). In the Middle Givet, the volcanism stopped, which resulted a peak of reef growth until the end of the Fasne/Famenne time boundary [5].

Cajus G. Diedrich

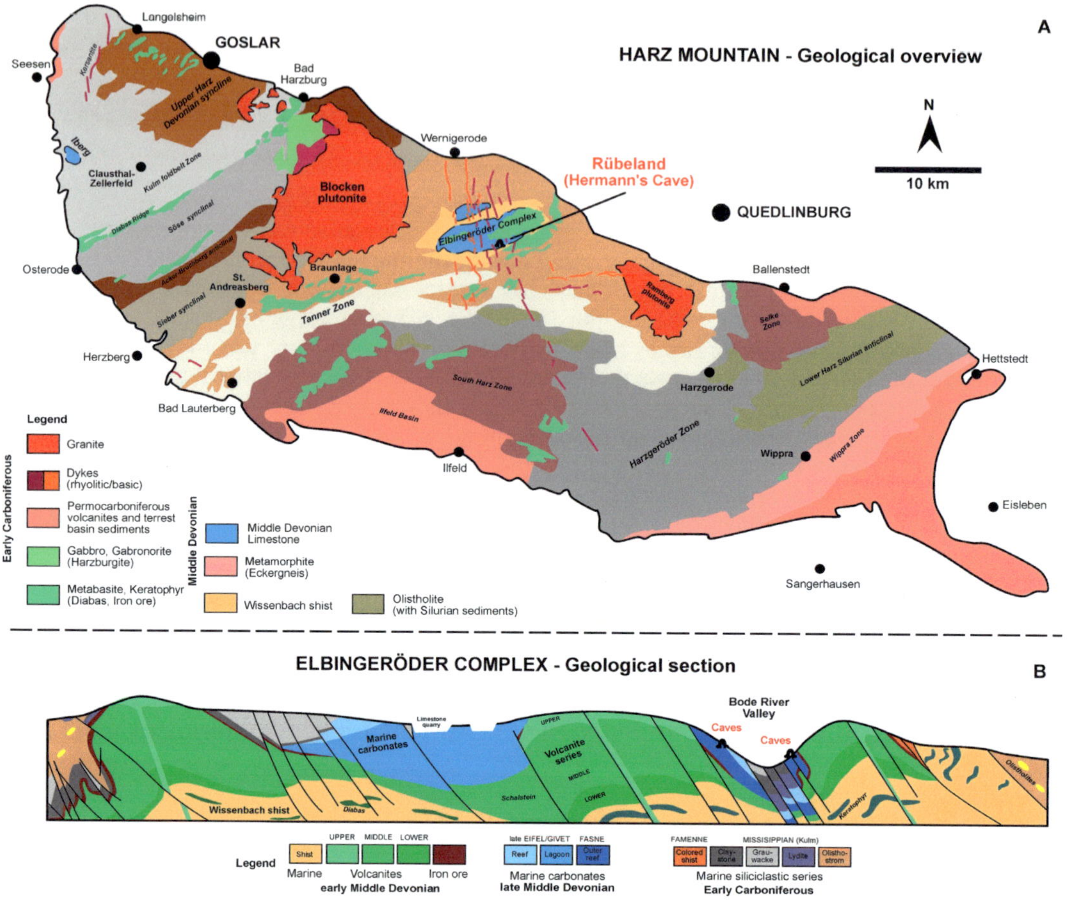

Figure 1. A. Geological overview of the Harz Mountain, and B. Karstic middle Devonian limestone area, the Elbingeröder Complex, in cross-section between the Büchen Mountain and Bode River Valley (redrawn, composed and modified from [1 - 5]).

In the Famenne, the marginal stromatopore reef lagoons disappeared and developed into the cephalopod lime mud facies (= outer reef facies limestone) and red claystones (= colored shist) [5]. During the latest Upper Devonian Dinantian, volcanic carbonates rose up in deep swell area clefts in submarine Neptunic dykes [5]. With the new marine transgression at the beginning of the Early Carboniferous (Missisippian), the siliciclastic Kulm facies developed which typical rocks are today Kulm-lydite, -claystone and -grauwacke (= grey-greenish silt/sandstone) or submarine tectonically initiated slump deposits of Olisthostroms with Olistholites [5] (Fig. **1B**). The Permian started at the beginning of the Rotliegend in a time of Central European mountain landscape terrest red bed depositions (fluvial conglomerate, sand and claystone), where south of the southern Harz Mountain famous vertebrate footprints and trackways were left along water bodies by smaller to medium-sized old reptile groups [6]. With the Zechstein marine transgression the intensive in and around the southwestern Lower Harz regions (Mansfeld, Sangerhausen/Eisleben *etc.*) mined fossil-rich Copper Shales (Fig. **1B**) [7, 8] were deposited in a hypersaline shallow Central European basin, where especially the rich fish fauna became well known [9], also from the Harz Mountain village Bad Lauterberg [8], which runs a small fossil museum (Heimatmuseum Bad Lauterberg) about local new unpublished finds. The Upper Permian Zechstein Basin evaporated finally, whereas first lagoonary dolomites and coastal oolite sands built a new carbonate series [10]. This is followed finally by gypsum and massive halite and other salt [10]. The gypsum surrounds in a belt the southern Harz Mountain margin. This area is another strongly karstified region [10], where Pleistocene sinkholes in Osterode have delivered Late Pleistocene hyena den bone accumulations [11]. In this Permian gypsum karst only a single larger and for the Bronze Age important cemetery cave is known with the Lichtenstein Cave [12]. In the Late Cretaceous (Santonian), the Harz block was uplifted by the subhercyn tectonic phase above the former sea level and became an "island" since the Middle Santonian [13] which continued in Tertiary times.

In Eocene to Oligocene times, dolines or sinkholes developed as first karstic structures which often filled with bean iron ore [14, 15], which depressions are often cut in limestone quarries nearby Rübeland. The larger valley-sided larger ponor caves in Rübeland (Hermann's or Baumann's) started to develop later, since the early Middle Pleistocene parallel to the valley glacier developments of the Mt. Brocken peak [16]. Those caves are key sites to understand the Harz glaciation process and landscape change development outside the cave [16] such as speleothem phases [17] in Pleistocene times. The caves contain relict sediment series, which are no more present in the Bode River Valley, or animal and archaeological remains that help to date those "cave deposits" and let reconstruct the climate and palaeoenvironment such as biodiversity [16]. There are three large known caves within the Middle Devonian limestones along the Bode River Valley in and nearby the Rübeland village: the Baumann's Cave, Hermann's Cave, and Brandes Cave [16]. Another fourth small cavity located downstream has important microfauna remains [16].

PLEISTOCENE CAVE GENESIS

Hermann's Cave Ponor Level Ages and their Refills

Underground water caused ponor caves [18] such as the Hermann's Cave (Fig. **2**) have typical fluvial underground water pressure caused small- to large-scaled facets along the limestone walls [16]. Examples are found in all elevations of the

Hermann's Cave. Those are best preserved with small facets in the lowermost still active ponor (Figs. **2, 3E**). In the Hermann's Cave, there are six ponor levels in total to distinguish: three larger Pleistocene and one Holocene elevation (Fig. **2**) [16] that eroded into the Middle Devonian Elbingeröder Formation reef limestones and its cleft systems. The three main Pleistocene cave genesis and ponor stages correlate to Middle and Late Pleistocene periods which are similarly distinguished for the Baumann's Cave [16] and the southern Harz slope Unicorn Cave Permian dolomites [16].

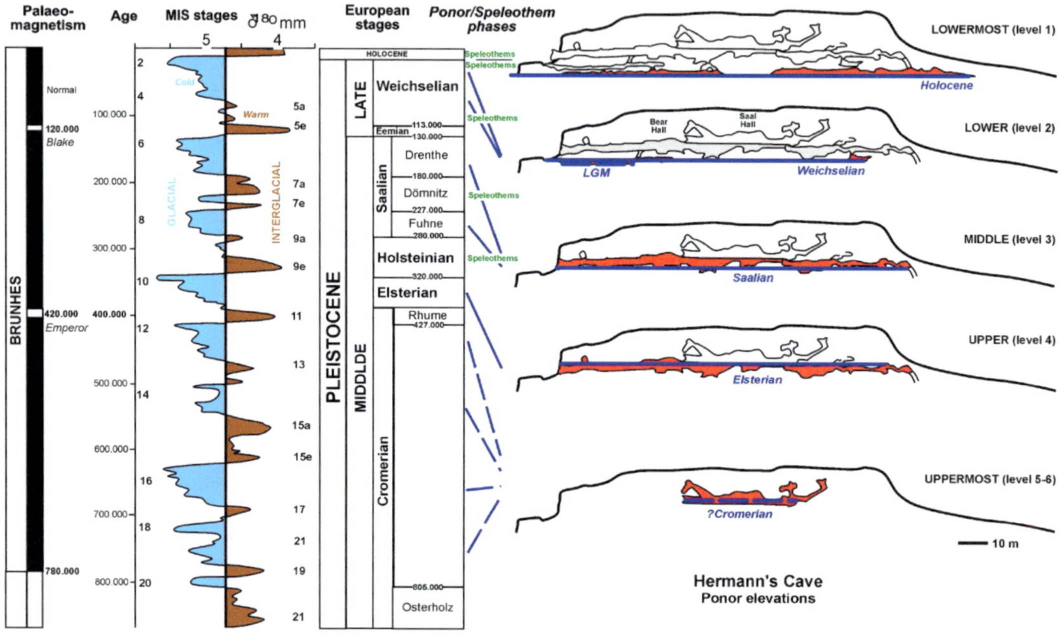

Figure 2. Climatic change caused development of the ponor levels of the Hermann's Cave. Those correlate to the glacial MIS phases. Speleothems formed mainly in Interglacials/Interstadial warm periods.

Uppermost Ponor Level 5 to 6 – Pliocene to Early Pleistocene

The Hermann's Cave must have started to form with dolines-short ponor structures first (Fig. **2**). Sediment samples from highest cave parts contain Bohnerz [16], an iron ore type of fluvial reworked origin, which is suggested to be of Tertiary origin. Other Tertiary filled sinkholes of the Rübeland karstic plateau contain similar ore enriched sediments [16, 17]. Absolute dating into the Late Tertiary Pliocene/Early Pleistocene of the uppermost levels is not yet possible due to the lack of datable relict sediments or fauna remains.

Upper Ponor Level 4 – Middle Pleistocene Cromerian to Elsterian

In the Hermann's Caves, the highest level 4 (Fig. **2**) forms a first long cave and is expected to represent an early Middle Pleistocene Cromerian (MIS 21 - 11) or Middle Pleistocene Elsterian (MIS 10) first underground ponor structure [16]. It must have resulted from the first still recognizable large glaciation of the Mt. Brocken and Harz Mountain [16]. One main Ice Age drainage was within the Bode River valley to the Northeast passing the Hermann's and Baumann's caves [16]. The Bode River ground water elevation drained several meters deeper underground through the forming caves which deposited fluvial sands and gravels especially during glacier melting periods [16]. To this age, the oldest ponor elevation of the Baumann's Cave and Unicorn Cave seem to correlate [16]. In the Unicorn Cave, Middle Pleistocene *U. deningeri* cave bears are believed to date into the Cromerian Complex (MIS 21 - 11) [cf. 16], or Holsteinian Interglacial (MIS 9). MIS 9 speleothem layers cover one bonebed layer in this cave [16], which date the oldest cave bears and cave den use indeed to be of Cromerian times [16]. In the Baumann's Cave cave bear bonebeds in one area seem also be of Cromerian or Elsterian Complex times [16]. In the Hermann's Cave, the Elsterian ponor elevation was not accessible because of the active ponor stream. Cave bears could not have used this cave as a den at that time.

Middle Ponor Level 3 – Late Middle Pleistocene Saalian

This level in the Hermann's Cave (similar in the Baumann's Cave) is dated in its formation by the Saal Chamber section (Figs. **2, 3**) [16]. Two meters thick fluvial sand/gravel sediments cover the cave floor below autochthonous Late Pleistocene cave clay and silt sediments [16] (Fig. **3**). Those fluvial series with their grauwacken, claystone and dark flat hornfels pebbles seem to originate from a glacial period - the late Saalian Glacial, or MIS 6, after the ponor cave had been formed in the MIS 8 - 7 times (Fig. **3**) [16]. The gravel and sand rock types of the Saalian Bode Valley glacier sediments originate from Mt. Brocken and the upstream bifurcating Bode River: Kalte and Warme Bode valleys [16] (Fig. **2B**).

It seems, another cave bear den period in another cave area of the opposite Baumann's Cave (old cave area) was used exactly in this Saalian Complex age [16]. The cave bear bonebeds in the Baumann' s Cave and the Unicorn Cave are mixed with magmatic Brocken rock types/local dolomites or limestones and finally both are covered by an older Saalian Complex aged speleothem layer [16, 19]. In the section of the Blasius Passage of the Old Cave part in the Baumann's Cave (correlating to the middle level 3 in the Hermann's Cave) teeth and bones of transitional European late *deningeri* to early spelaeoid forms [16] were found below a thick speleothem layer, that is dated by Uranium isotopes 236.340 ± 16.429 or 232.474 ± 40.141 BP [17]. Those data fall into the MIS 7 Interglacial period [16] and support the forming age of the middle ponor level 3 of the

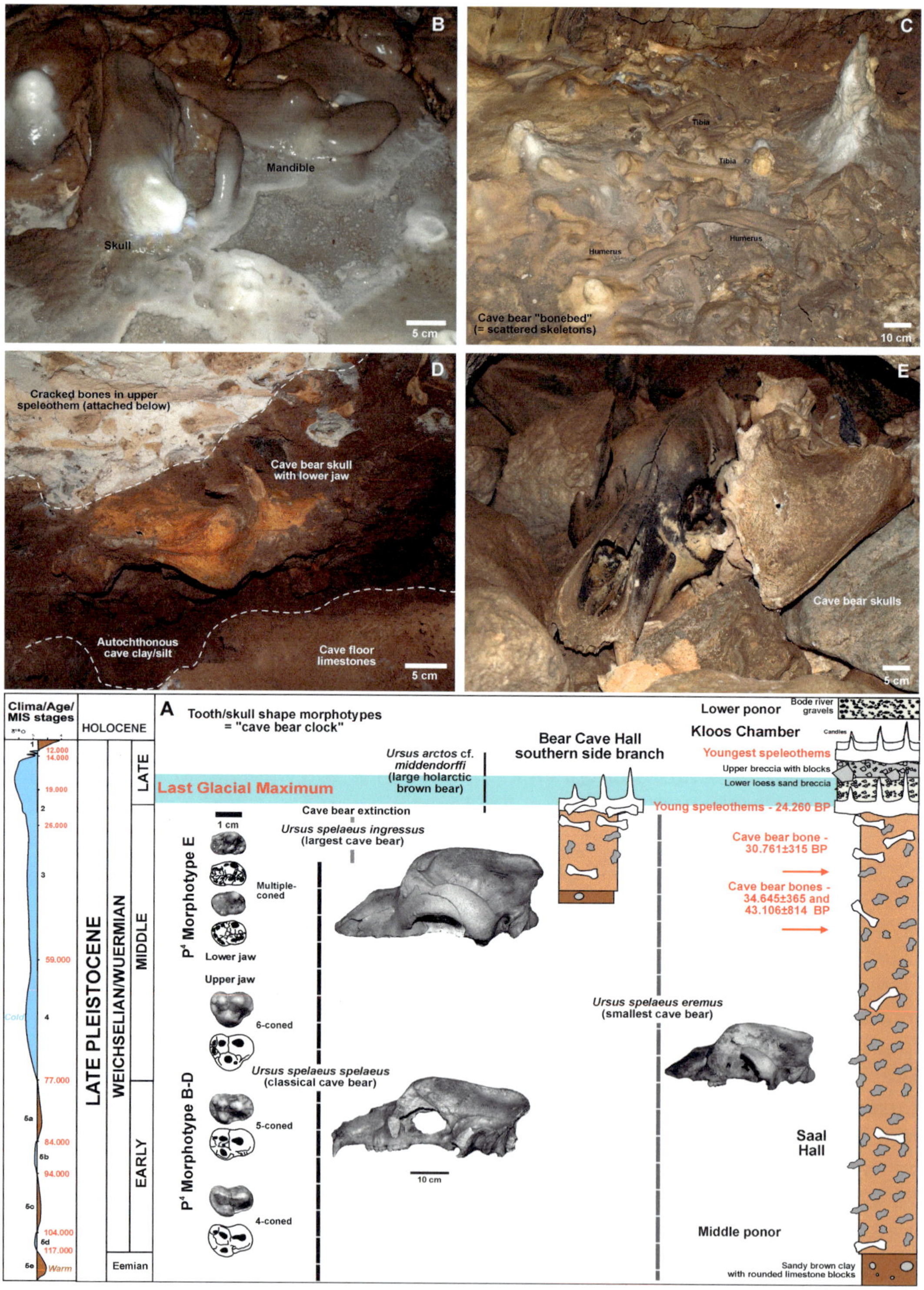
B
Mandible
Skull
5 cm
C
Tibia
Tibia
Humerus
Humerus
Cave bear "bonebed"
(= scattered skeletons)
10 cm
D
Cracked bones in upper speleothem (attached below)
Cave bear skull with lower jaw
Autochthonous cave clay/silt
Cave floor limestones
5 cm
E
Cave bear skulls
5 cm
Clima/Age/ MIS stages
HOLOCENE
12.000
14.000
19.000
26.000
59.000
Cold
77.000
84.000
94.000
104.000
117.000
Warm
LATE PLEISTOCENE
WEICHSELIAN/WUERMIAN
LATE
MIDDLE
EARLY
Eemian
A
Tooth/skull shape morphotypes = "cave bear clock"
Last Glacial Maximum
Ursus arctos cf. middendorffi (large holarctic brown bear)
Cave bear extinction
1 cm
Ursus spelaeus ingressus (largest cave bear)
P4 Morphotype E
Multiple-coned
Lower jaw
Upper jaw
6-coned
Ursus spelaeus spelaeus (classical cave bear)
P4 Morphotype B-D
5-coned
4-coned
10 cm
Bear Cave Hall southern side branch
Ursus spelaeus eremus (smallest cave bear)
Lower ponor
Bode river gravels
Kloos Chamber
Candles
Youngest speleothems
Upper breccia with blocks
Lower loess sand breccia
Young speleothems - 24.260 BP
Cave bear bone - 30.761±315 BP
Cave bear bones - 34.645±365 and 43.106±814 BP
Saal Hall
Middle ponor
Sandy brown clay with rounded limestone blocks

Figure 3. A. Taphonomic situation of the cave bears in the Hermann's Cave. A. Generalized stratigraphy of the middle ponor level 3 and absolute dates on speleothems and of the Hermann's Cave. Cave bear species/subspecies and indirect dating by P4 morphology = "cave bear clock" (composed and modified from [16], MIS stages and climate curve from [16]). B. Skull of the large cave bear *Ursus ingressus* and besides one mandible at the end of the Saal Hall covered by the youngest speleothem layer. This one seems to be within a round hibernation nest area, the last at the end of the cave 90 m far from the former entrance. C. The floor of the Bear Hall was in former times covered like this relict bonebed area. Similar sized humeri and tibiae and a not far in-situ skull are again from a large *U. ingressus,* possibly one individual, whereas other bones around (*e.g.* visible vertebrae, *etc.*) also might belong to that individual, which surface was also covered and preserved by the latest speleothem generation. D. Whereas bones are sticking out of the speleothem surface, those are also cemented into the Post-LGM dated speleothem layer. Below, most of the cave bear skulls (here an articulated skull with its lower jaw – again prove for autochthonous situation) and other bones were excavated in the Bear Hall. This skull is at the end of a small side tunnel, where Steiner excavated 1964. E. A nearly complete skull of a young "cave bear" (probably *Ursus spelaeus eremus*) and another braincase, which along with other scattered bones, were not covered by neither speleothems nor sediment or fallen blocks. Those are still preserved hidden in a niche between the bear and Saal Hall.

Hermann's Cave during the late Saalian Complex MIS 7, possibly even continuing in the Eemian Interglacial MIS 5e.

The second refill phase of the level 3 happened in the early-middle Late Pleistocene (MIS 5d-3). In the Hermann's Cave no floods went over the skeletons/bones, so those are in-situ or more or less autochthonous in their position within red cave clays and silts which look like bonebeds (Figs. **3B-D**) [16]. Although surface excavation documentation is missing in any former publications at some places the presence of cave bear skeletons on-place can be demonstrated (see Chapter cave bears). Often skulls are found besides or in articulation with their lower jaws (Figs. **3B and D**) [16]. The problem is the severe damage of many of those skeletons and "bone-rich areas" (or bonebeds) due to trophy hunting of the skulls and mandibles up to recent times. Articulated skeletons, which were covered by speleothems in the Bear Hall (Fig. **3C**) and at some places articulated legs or vertebral columns were present. Those indicate bears died either in their hibernation nests or at least within the cave [16]. First absolute dating of the red clays and silts were made on cave bear bones from the Bear Cave Hall and Saal Hall, which are non-calibrated old dating, and somehow unclear in the stratigraphy with Late Pleistocene age 30.700 years ago (calibrated around 32.700 BP = MIS 3) [14], support the relative dating. Other cave bear bones (samples from the bone dump of the Hermann's Cave [cf. 16]) are AMS 14C-dated between 34.645±365 and 43.106±814 BP [18] and do not represent the age of all cave bears nor the fauna in general. Furthermore, the younger speleothems covering the bone layer have a maximum age of 25.000 BP [17]. Similarly the cave bears of the upper layers in the Unicorn Cave date by P4 cave bear tooth morphology into the early/middle Weichselian [16]. The P4-teeth from this cave are mainly B-D-types with three to four main cones (lower layers) and E-types = latest forms with multiple enamel surface cones (uppermost layer) [16].

Figure 4. Main different sediment types in different ponor levels of the Hermann's Cave. A-B. The youngest LGM/Post-LGM frost breccias cover most of the middle ponor cave system, which themselves are cemented in larger areas by the younger speleothems, especially with typical candle stalagmites. C-D. Into this younger speleothem layer at many places the cave bear bones are cemented. Below there are the autochthonous cave clays-silts in the middle Saalian formed ponor level (refilled during Weichselian) which have only smallest local Devonian limestone pebbles and bones including micromammals. E-F. In the lowermost active ponor Holocene floods from the Bode Valley are pressing typical F. river terrace gravels with Granite, Grauwacke or Lydite which originate outside the cave from the Bode River upstream to the Brocken peak leaving thereby faceted limestone walls (composed and modified from [16]).

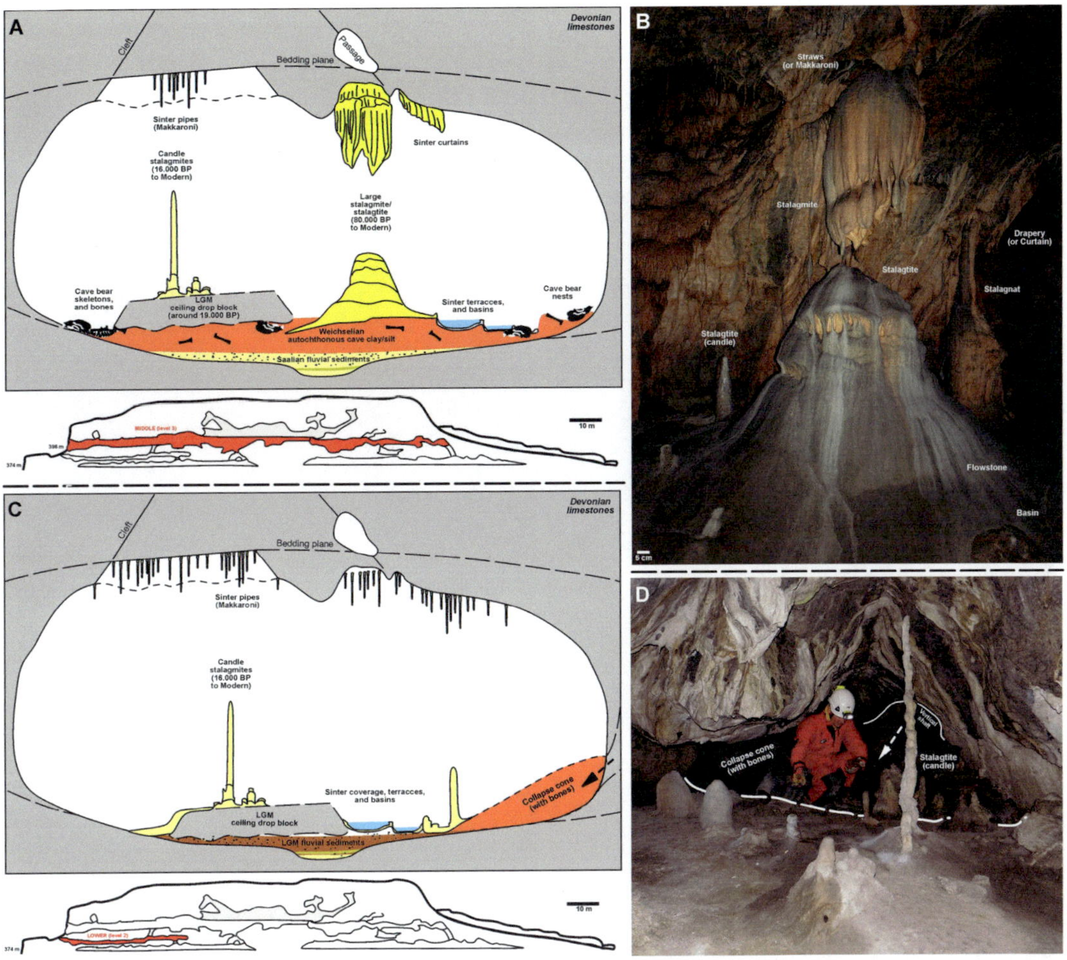

Figure 5. Main speleothem types of different speleothem genesis times and elevation levels in the Hermann's Cave. A-B. Older Early-Middle Late Pleistocene (MIS 5a and c Weichselian Interstadials) speleothems in the Bear Hall of level 3. C-D. Smaller post-LGM (MIS 2-1) speleothems with typical long candles in the Lower Schwemm Passage of level 2. Those cover also through vertical cones down to 8 meters collapsed bonebed layer material on its corners.

In the Hermann's Cave, the cave bear giants of *Ursus ingressus* (see Chapter Cave Bears) are found all over the cave floor of level 3 (also partly level 4) (Fig. **3**). These last giant cave bears of Europe are AMS 14C-dated in southern German caves between 32-25.000 BP [19]. This falls also into the time of the European Aurignacian/Gravettian Late Palaolithic Cromagnon humans, in which few artifacts were found below the speleothem layer in the upper cave clays-silts (see Chapter Aurignacians).

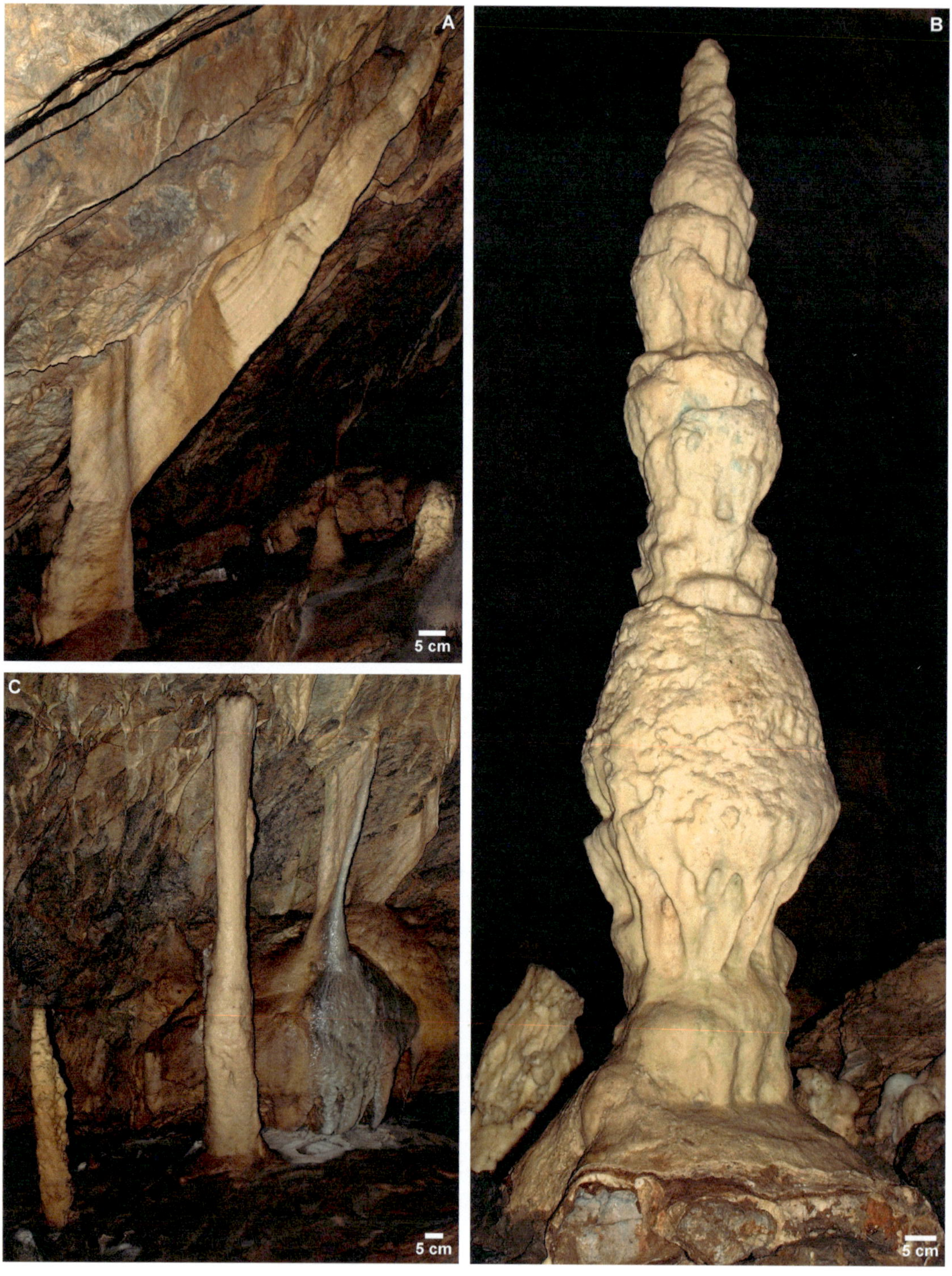
A
5 cm
B
5 cm
C
5 cm

Figure 6. Older main different speleothem types in the Middle Ponor level (Saalian Complex origin and refill with Late Pleistocene speleothem decoration) of the Bear Cave Hall, Hermann's Cave. A. *Sinter curtain* which developed along a cleft. B. *Stalactite* produced of splashing water drops in cascade form. C. *Stalagnate* (middle) connecting the floor and ceiling (left side candle stalactite).

Lower Ponor Level 2 –LGM

The deeper final Late Pleistocene cave ponor system was formed (Fig. **2A**) starting after 25.000 BP with its peak in in the Post-LGM due to seasonally melting glaciers and complete melting within the Dryas to Alleröd climate changing periods (final MIS 2) [16]. In the overlaying levels other sediments accumulated especially within the LGM. Large ceiling blocks and frost breccias are found especially in the former entrance area of level 3, but in all chambers and halls up to several meters thick breccias cover the cave bear bonebeds (Figs. **3A-F**, **4A**) [16]. Those or at places where no breccias formed, the youngest speleothem layer covers those including typical candle stalagmites (Figs. **4A**) [16].

Lowermost Ponor Level 1 – Holocene

It remains as today's Bode River valley ground water drainage (Fig. **2A**), which is active during water high stands about one to two meters below the recent river terrace, which ponor stream in the cave falls dry in less humid periods [16]. Small corrosion facets ornament the limestone walls (Fig. **4E**) [16]. The fluvial sediments are gravels, which are similar as those of the Bode River terrace, but those exclude larger pebbles Its components are mainly grauwacke, shist, lydite and smaller granite pebbles (Fig. **4F**) [16]. Important is the absence of candle speleothems supporting its youngest age after the last Post-LGM speleothem phase. Another find for indirect dating both levels 1 and 2 is a brown bear skeleton remain of the Lower Fluvial Passage of level 2 of the Hermann's Cave (see Chapter LGM) that is in bone preservation less fossilized, and yellowish and different in preservation to the cave bear bones of the upper levels 3 and 4. These remains were found somewhere nearby the second former entrance on the river gravels dating the fluvial and latest speleothem phase covered sediments additionally into a LGM-Post-LGM period (Fig. **3E**) [16].

Hermann's Cave Speleothems

Speleothem Forming Periods

Holsteinian Interglacial (MIS 9) and Saalian Complex Interglacial (MIS 7) speleothems are not securely present in the upper levels 5-6 of the Hermann's Cave, but built the large and massive speleothem decorations within the nearby

Figure 7. Younger speleothems (mainly candles) in the Lower Ponor level (LGM-Post-LGM origin and refill with Latest Late Pleistocene/Holocene speleothem decoration) of the Lower Schwemm Passage, Hermann's Cave.

Baumann's Cave and Unicorn Cave [16, 17]. In the Hermann's Cave four speleothem genesis phases can be distinguished all dating into the Late Pleistocene and its interstadials or glacial/interglacial boundaries [16, 17]. The first speleothems started to form and has continued in some cases until today in the two early Weichselian Interstadials of the MIS 5a = 104-94.000 BP, and MIS 5c = 84-77.000 BP [16, 17] (Figs. **3E**, **5A**, **6**). The third speleothem period falls within the interstadials of the Dryas to Alleröd times at the end of the MIS 2 = 16-14.000 BP where smaller and long candles started to grow until today [16, 17].

Speleothem Types

The typical and main speleothems (see definitions in [20]) are present in the Hermann's Cave. *Macaroni* are preserved at many places on the ceilings of the non-active ponor levels 2-6 which are in some cases at the beginning of the stalagtite development (Fig. **5B**). Such pipes where the water drops produce small growth rings at the end of the tube sometimes became blocked forming eccentrics.

Sinter basins are preserved with small-scaled ones in level 3 and 2, *e.g.* in the Bear Hall (Figs. **5A-B**), but often fall dry. *Sinter curtains* (definitions cf [20]) developed especially along clefts (Figs. **5B**, **7A**). *Cascade stalactites* were produced by splashing water drops (Figs. **5B**, **6B**). *Stalagmites* connect the floor stalagmites and ceiling stalagtites (Fig. **6C**). Thin *candle-like stalagmites*, especially up to several meters high, are preserved in the lower ponor levels 3 and 2 (Prae-LGM refill plus LGM speleothem decoration) and are found especially well preserved at the beginning of the visitor loop of the Hermann's Cave in the Lower Schwemm Passage (Fig. **7**). There are many other small speleothem forms such as excentriques or cave pearl types, which latter are exposed with some examples in the visitor entrance center of the opposite Baumann's Cave.

REFERENCES

[1] Liessmann W. Historischer Bergbau im Harz. 3rd ed. Berlin, Heidelberg: Springer 2010; p. 470. [http://dx.doi.org/10.1007/978-3-540-31328-1]

[2] http://www.lagb.sachsen-anhalt.de/ Landesamt für Geologie und Bergwesen

[3] Weller H. Eine neue Geologische Karte des Elbingeröder Komplexes/Harz - GK EK 2011. Hall Jb Beih 2011; 28: 57-62.

[4] Stedingk K, Rappsilber I. Geologisch-montanhistorische Karte Elbingeröder Komplex 1 : 20 000 mit Geologischer Karte Elbingeröder Komplex 1 : 25 000: Geotourismus im Mittelharz GMK 20 Elbingerode Landkarte - Folded Map, 2014.

[5] Weller H. Facies and development of the Devonian (Givetian/Frasnian) Elbingerode reef complex in the Harz Area (Germany). Facies 1991; 25: 1-50. [http://dx.doi.org/10.1007/BF02536754]

[6] Voigt S Tetrapodenfährten im Rotliegend. Schr dt Ges Geowissensch 2012; 61: 161-75.

[7] Knitzschke G, Spilker M. Die Kupferschieferlagerstätte Mansfeld/Sangerhausen - Bergbauliche Nutzung und Verwahrung. Der Anschnitt 55 2003; 3-5: 134-47.

[8] Wolfgang L. Der Bergbau auf Kupferschiefer. Wilfried Ließmann, Kupfererzbergbau und Wasserwirtschaft, Zur Montangeschichte von Bad Lauterberg/Südwestharz. Duderstadt 2001; p. 261.

[9] Diedrich C. Taphonomy and palaeoenvironment of the coelacanthid rich site Hasbergen (NW Germany) - a first systematic excavation in the Copper Shales (Upper Permian) of Europe. Palaeodiv Palaeoenvironm 2009; 89: 67-94. [http://dx.doi.org/10.1007/s12549-009-0004-6]

[10] Hermann A. Einführung in die Geologie, Morphologie und Hydrologie des Gipskarstgebietes am südwestlichen Harzrand. Jb Karst- Höhlenk 1969; 9: 1-10.

[11] Diedrich C. Seltene Freilandfunde der späteiszeitlichen Fleckenhyäne *Crocuta crocuta spelaea* (Goldfuss 1823) in Sachsen-Anhalt und Beitrag zu Horsttypen der letzten Hyänen im Jung-Pleistozän von Mitteldeutschland. Z Mitteld Vorgesch 2014; 94: 1-19.

[12] Flindt S. Die Menschen aus der Lichtensteinhöhle. Epoc 2010; 5: 74-81.

[13] Diedrich C, Scheer U. Selachian, teleostean, and reptile remains from the Upper Santonian (Upper Cretaceous) of Ochtrup (north-west Germany) and their importance for the palaeogeography of the southern Proto-Northsea Basin. Open Geosci 2015; 7(1): 2391-5447.

[14] Gross V. Stoffliche Analyse des Sedimentmaterials im Korngrößenbereich 0.06-20 mm der Hermannshöhle (Rübeland, Harz). Unpublished Diploma Thesis, Freie Universität Berlin 1977; 49.

[15] König W. Zeitliche und genetische Einordnung von tertiären Sedimentvorkommen im Mittelharz und Harzvorland – ein Beitrag zur Reliefentwicklung und zur Karstmorphogenese im Harz. Unpubl Thesis, Martin-Luther-Universität Halle-Wittenberg Institut für Geowissenschaften Physische Geographie 2005.151

[16] Diedrich C. Impact of the German Harz Mountain Weichselian ice-shield and valley glacier development onto Palaeolithics and megafauna disappearance. Quat Sci Rev 2013; 82: 167-98. [http://dx.doi.org/10.1016/j.quascirev.2013.08.013]

[17] Nordhoff P. Stable isotope investigations on speleothems from different cave systems in Germany. Unpublished Dissertation thesis, Georg-August-University Göttingen 2005; 149.

[18] Rosendahl W, Döppes D, Joger U, *et al.* New Radiometric datings of different cave bear sites in Germany - results and interpretations. Bull Soc Hist Nat Toulouse 2005; 141(1): 39-46.

[19] Münzel SC, Stiller M, Hofreiter M, Mittnik A, Conard NJ, Bocherens H. Pleistocene bears in the Swabian Jura (Germany): Genetic replacement, ecological displacement, extinctions and survival. Quat Int 2011; 245: 225-37. [http://dx.doi.org/10.1016/j.quaint.2011.03.060]

[20] Moore GW, Sullivan N. Speleology: Caves and the Cave Environment. Cave Books 1997; p. 176.

CHAPTER 3

Small and Large Cave Bears

Abstract: New field work and studies on 4.910 "cave bear" bones from one of the largest European cave bear dens, the Hermann's Cave of Rübeland in the Harz Mountains of Saxony-Anhalt in northern Germany, allow the reconstruction of the cave use in different cave areas being limited on the two upper ponor levels by three different cave bear species/subspecies of the Late Pleistocene time, when those areas were already dry. Cave bear polish along many parts on the limestone walls are documented mainly in the Bear Hall area. Partly articulated skeleton remains there and all over the Saalian created ponor level, such as autochthonous bonebeds, indicate non-fluvial transport of bones, because the active Weichselian cave ponor river stream drained at that time already 8-10 meters below. A neonate or Stillborn such as a sibling cave bear skeleton from the Hermann's Cave Bear and Saal halls and single bones and deciduous teeth are quite abundant in the sediments. The sections and bonebeds are partly dated absolutely with some C14 data of cave bear teeth ranging between 30.761-43.100 BP. The youngest Uranium dated 24.260 BP old speleothems of the beginning of the climatic change to the Last Glacial Maximum (LGM) cover the bonebed in the Bear Hall, which fits to the time of their extinction in Europe around 24.000 BP. Further dating is possible relative-chronologically using cave bear premolar P4 tooth morphotypes C-E and herein new classified three skull shape morphology types A-C in combination with the bone preservation and new section studies. The cave bear den time started most probably already in the Eemian Interglacial MIS 5e, but it was in use mainly in the early-middle Weichselian or Wuermian Glacial (MIS 3-5d). The longbone proportions support the separation between three different sized species" 1. The smallest cave bear *Ursus spelaeus eremus* (skull shape B, P4 morphotype C), 2. The medium-sized *Ursus spelaeus spelaeus* (skull shape C, P4 morphotype D), and 3. The largest cave bear *Ursus spelaeus ingressus* (skull shape D, P4 morphotype E) of the MIS 3 time. Most probably, *U. s. spelaeus* evolved during the Late Pleistocene into t *U. s. ingressus*, whereas the smallest *U. s. eremus* lived with both sympatric all the time in middle high elevated non-alpine European boreal forest mountain regions. Compared to northern American bears, the small cave bears *U. s. eremus* lived possibly similar as extant black bears, the *U. s. spelaeus* forms more similar but much more herbivorous as extant grizzly brown bears. *U. s. ingressus* was full herbivore specialized with its multiple-coned and enlarged tooth cusps. Palaeopathologies on teeth, jaws and postcranial bones are found on all cave bear types and at different age classes. Some bone injuries resulted into trauma bite damage which were caused by other bears or large predators (mainly their canines) from inter- or intraspecies fights. The mortality of cave bears, especially siblings and cubs, is strongly influenced by top predators, Ice Age steppe lions *Panthera leo spelaea*, Ice Age spotted hyenas *Crocuta crocuta spelaea* and Ice Age wolves *Canis lupus spelaeus*. Hyenas and wolves and in

Cajus G. Diedrich

much lesser amount to prove lions are the main bite mark and bone damage producers of many cave bear bones. About 80% of the bear siblings and cub bones (3 months to one year in individual ages) have chewed longbone joints or expose bite damage, but are much lesser cracked due to the still soft compacta. Bones of subadults to elderly are only up to 20% large carnivore damaged. Cave bear cubs and siblings were either killed or at least only scavenged by all three top predators, whereas the cub hunting specialization by steppe lions is well known due to nitrogen isotope signs. Chewed/damaged bones in similar bite damage stages classified into three stages including pseudobone flutes with hyena bite impact holes in cave bear cub femorae are found not only in the Hermann's Cave and are known from many other cave bear dens all over Europe. Cave bears of the middle high elevated mountain Boreal forests regions like in the Harz Mountain were the main prey of lions, hyenas and wolves during the Late Pleistocene, as result of the absence of the mammoth steppe big game. Mammoth, rhinoceros, bison or mainly reindeer herds such as Cromagnon human Aurignacien hunters were migrating seasonally from the Harz Mountain Fore Lowlands of northern Germany upstream along the river valleys, such as the Bode Valley, reaching thereby the Hermann's Cave for cave bear hunting deeper in the cave, which is supported in the Bear Hall and Saal Hall.

Keywords: Late Pleistocene cave bear den, Three cave bear species/subspecies, Taphonomy, Palaeoecology and pathology, Predators and scavengers, Hermann's Cave, Harz Mountains of northern Germany (central Europe).

CAVE BEAR BONE EXCAVATIONS

The first cave bear skulls, teeth and bones in the Hermann's Cave were found directly with its discovery during road construction works and excavations were started by Grotrian 1878 [1], an later by excavations of Kloos and Müller in 1889, especially in the Bear Hall area [2], which bones were dumped there in part for visitors already in historic times (Figs. **1A**). At that time, the bone layers and even skeletons were removed typically by the use of pick-axe and shovels, especially the larger material. Kloos composed from this first excavated material from different animals and cave bear species a first skeleton [2], which had been exposed for a long-time within the Bear Hall (Figs. **1B**, **2**).

The original skeleton (Figs. **1B**, **2A**) was sadly uninstalled and thrown onto the cave bear bone dump being replaced with a replica skeleton in 2010, which is still on display. 2011 the author of this book has saved and restored the original skeleton (Fig. **2B**) which will be used in a new presentation in future [3]. A second skeleton was mounted by Kloos for the former "Naturhistorische Museum Braunschweig" which was some years ago new mounted (Fig. **3**) in the new named Staatliche Naturhistorische Museum Braunschweig [3].

Those two composite "cave bear" skeletons were produced in historic times including the medium-sized and larger cave bear species or subspecies as being discussed herein. Because of their historic value, they were left as they were mounted with the historically used bones and will not be rebuilt as correct cave bear species composite skeletons in future. Other bones are partly not in correct position, similar sizes or proportions, or are missing. In both skeletons the sterna bones are missing or were not included (Figs. **2B**, **3**). Especially, in all mounted "cave bear" skeletons over Europe, the nine tongue bones are absent, which were composed for the first time in a non-mounted exposed skeleton of the Sophie's Cave small cave bear subspecies of *Ursus spelaeus eremus* [4]. Typically, the pedal skeletons have no small sesame bones included, such as seen in the skeletons which were mounted by Kloos. The last larger excavation in the Hermann's Cave was made 1985 by Völker and the speleoclub Harz in the Saal Hall [5] (Fig. **1C**), which delivered 2.230 bones [5, 6]. Most of them are cave bear remains, but the material includes several steppe lion bones (see Chapter Steppe lions).

Figure 1. A. Historical situation of the Bear Hall bone dump. B. Already in the 70's partly damaged original skeleton which was mounted by Kloos (cf. Fig. **2A**) in the Bear Hall of the Hermann's Cave. C. Völker excavation 1985 in the Saal Hall.

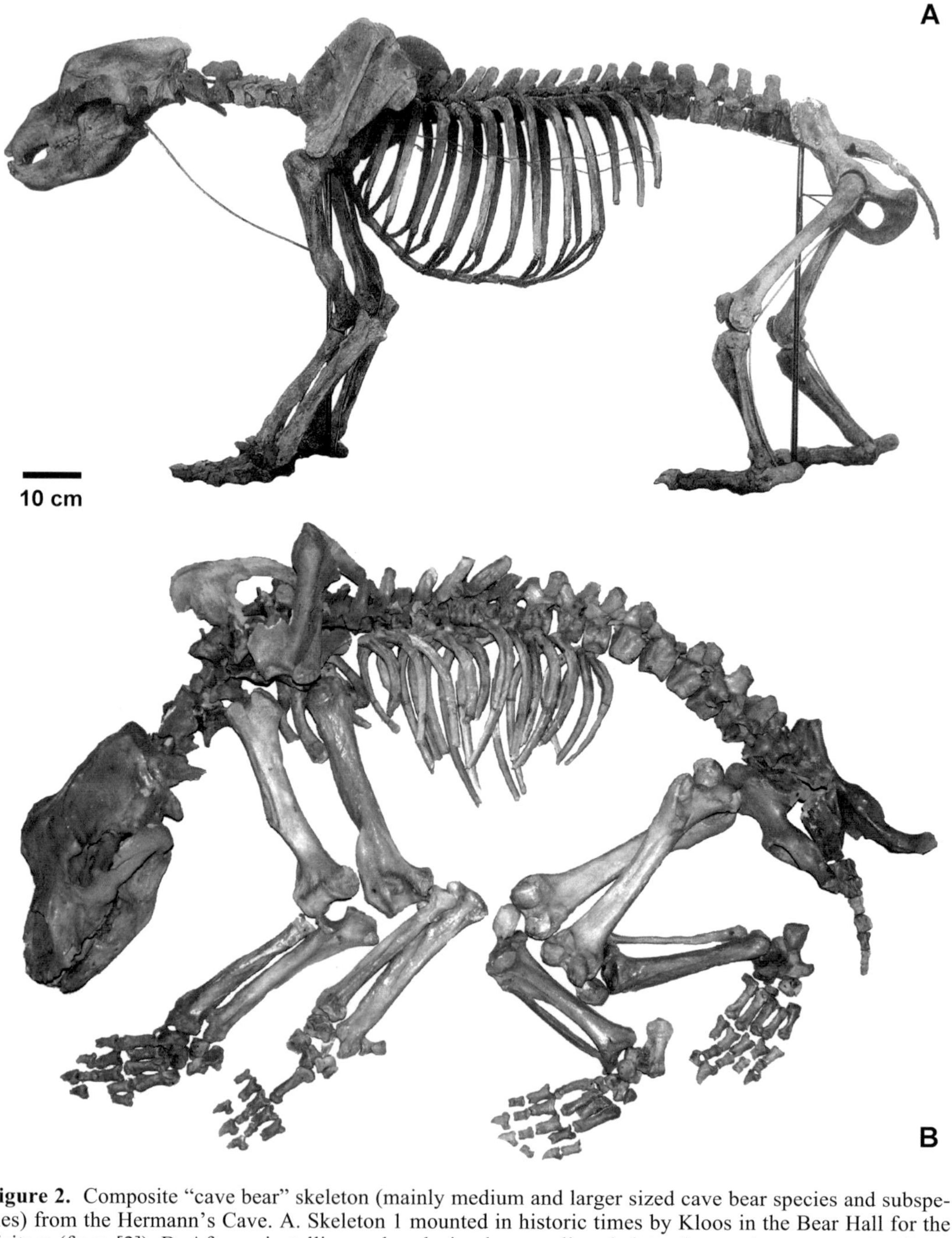

Figure 2. Composite "cave bear" skeleton (mainly medium and larger sized cave bear species and subspecies) from the Hermann's Cave. A. Skeleton 1 mounted in historic times by Kloos in the Bear Hall for the visitors (from [2]). B. After uninstalling and replacing by a replica skeleton it was thrown onto the "bone dump". It was saved in 2011 for restoration whereas some of the pedal bones and rib parts bones are lost (coll. RC).

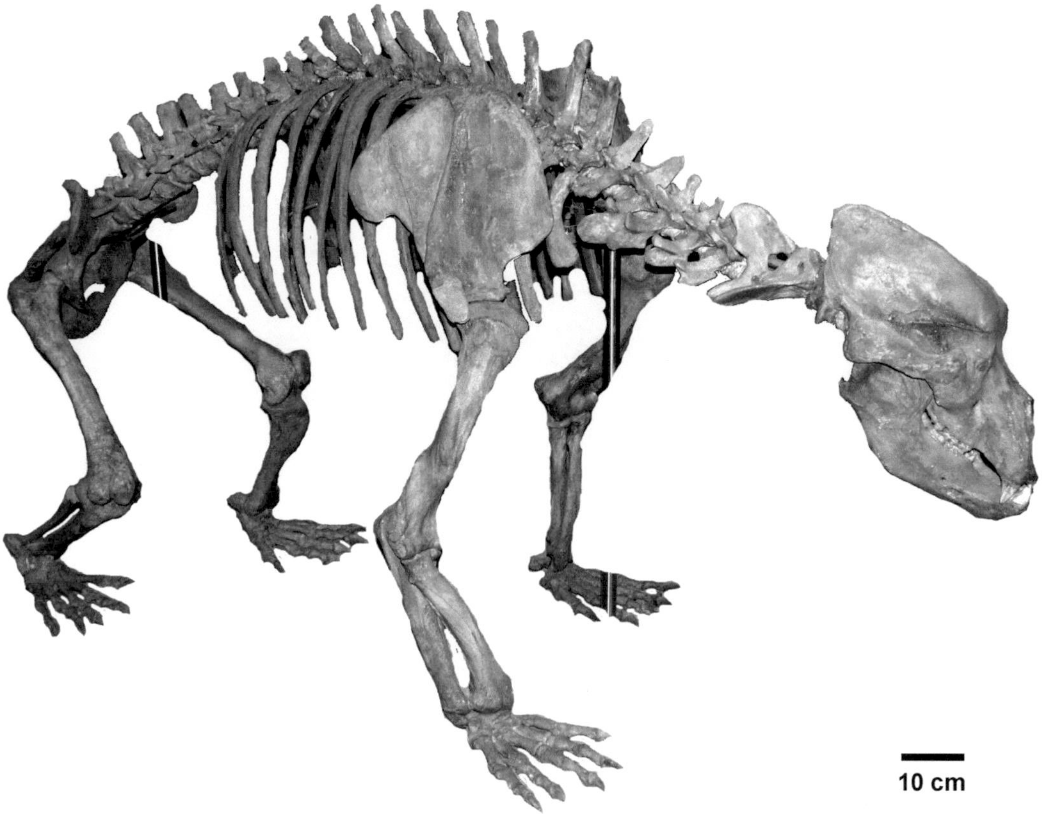

Figure 3. Composite „cave bear" skeleton (mainly medium and larger sized cave bear species and subspecies *Ursus spelaeus spelaeus* and *Ursus spelaeus ingressus*) from the "Rübeland Caves" (mainly, or all bones from Hermann's Cave) after its restoration. It is the second skeleton mounted in historic times by Kloos (coll. LMNB).

The Large Cave Bear Den

The cave bear den area within the four main horizontal ponor leveled Hermann's Cave system (Fig. **4**) was mapped in 2011 [3, 7] using a. The distribution of bones partly being encrusted by speleothems on the cave floor and in its sediments, bones being exposed in sections (Figs. **5**, **6**), B. Cave bear polished walls which were discovered in 2011 (Fig. **7**) [3, 7], and C. Cave bear hibernation nests. Cave bear bones and cave bear polished walls such as hibernation bed depressions (Fig. **8**) and more rare footprints or scratch marks in cave clay or scratches at softer limestone walls (Fig. **5**) are the typical important indicators for cave bear den identifications [8].

The Bear, Collapse and Saal halls and horizontal connecting passage levels of the Hermann's Cave (Fig. **5** – red area) were formed in the late Middle Pleistocene

Saalian Glacial (see chapter Geology and Sedimentology) [7]. Only those upper levels were used in a length of 90 meters from the former entrance by all cave bear species or subspecies at a time that this ponor level was already the dry cave section during the Late Pleistocene [3, 7].

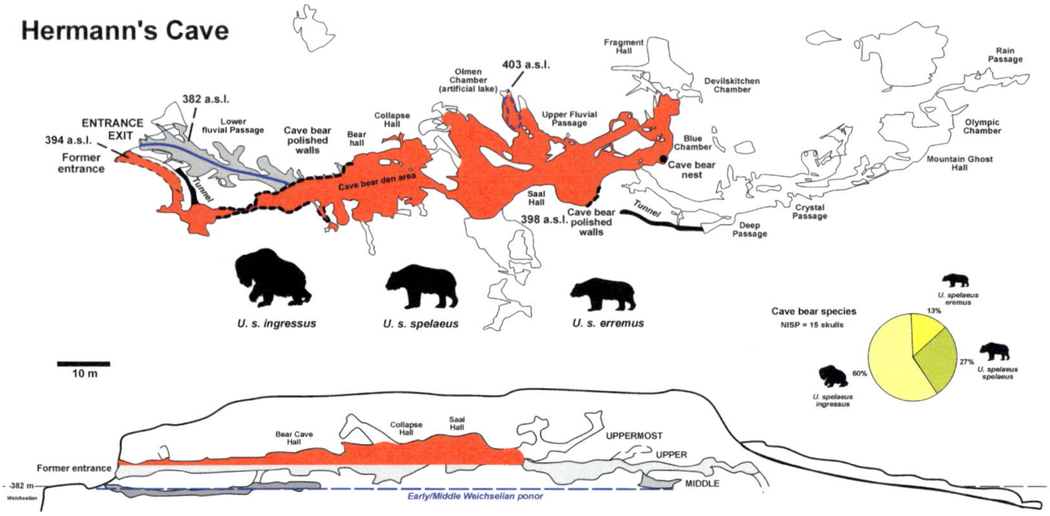

Figure 4. Hermann's Cave cave bear den map. A. The cave bear den area (red) reaches up to 90 meters far from the former entrance over the Saal Hall to the Blue Chamber. During the Late Pleistocene, the active ponor stream was 8-10 meters below the dry cave bear den area of the former Saalian ponor level. Statistics of the cave bear species based only on skulls.

Skeletons, Skulls and Bone Distribution

Recently, only relict areas are preserved with in-situ bonebed areas which are now quite limited (Fig. **5**). Those survived in areas where the LGM speleothem layer cemented those (Figs. **5A, B**) or in hidden corners (Figs. **5D-E**), such as under LGM frost breccias gravels [3]. The floor of the Bear Hall in former times was covered like the figured relict bonebed area (Figs. **5A**). At many places, where skulls and bones are still on the surface and uncovered by speleothems, those are either trampled by modern humans (spelunkers) or are partly weathered nearly to dust (Fig. **5C**).

Cave bear skulls are in-situ and in some cases in articulation or close contact to their lower jaws (Figs. **5B, C, E**) at several places in the hibernation area of the Saalian ponor level. This is present in the Blue Chamber with a skull of *Ursus spelaeus ingressus* (Fig. **5B**), suggesting again death and embedding of subadult to elderly cave bear carcasses on place. A nearly complete skull of a subadult "cave bear" (probably *Ursus spelaeus eremus*) and another braincase, which were not covered with speleothems nor sediments or fallen blocks are preserved hidden in a niche between the Bear Hall and Saal Hall together with three cub foreleg bones (Figs. **5D-E**).

Trampling of skeletons and bones by cave bears, carnivore predator scavenging activities (lions, hyenas and wolves – see Chapters Lions, or hyenas and wolves) and Late Aurignacien humans (see Chapter Archaeology) in most cases seemed to have caused only short-distance lateral dislocation of several carcass skeletons of adult to elderly individuals. The articulated incomplete neonate or sibling cave bear skeletons from the Saal Hall prove well the autochthonous character of the cave bear "bonebeds" [3]. Unfortunately, at many niches of the cave and along the walls, skeletons have been destroyed or heavily damaged in the past 100 years by cavers and visitor cave work such as several installations of electric cables in the cave floor sediments (Fig. **5C**). Similar in size a humeri and tibiae and not a far distance away a in-situ skull (Fig. **6**) are from the large *U. spelaeus ingressus*, most probably representing a single individual due to its similar bone proportions. Other bones around (*e.g.* vertebrae, *etc.*) also might belong to that individual, which has a large bite mark and diagonal cut on one proximal humerus made by a top predator (lion or hyena). The skeleton was covered and cemented by the last speleothem generation (Figs. **5**, **6**). Connected vertebral column parts, legs, or possibly complete articulated skeletons are located not only in the Bear Hall. Their excavations would have to use surface excavation technique in the future in able to understand the taphonomy of skeletons, or the presence of skeletons within hibernation nests.

Cave Bear Polish at Limestone Walls and Exposed Corners

Cave bear generations polished in the Hermann's Cave over ten thousands of years at several places the limestone walls and its exposed corners, especially in the narrow passages in the last stage highest elevations about one meter above the cave floor [3] (Figs. **4**, **7**). The best preserved and most polished cave limestone walls are found in the narrow passage between the former entrance and the Bear Hall. Further polish is preserved in the Saal Hall (Figs. **4**). Such polished limestone walls or corners are common in many larger German and European cave bear dens [9, 10]. Cave bears polished narrow passages by chance when entering and exiting the cave for hibernation. They furthermore must have polished again and again the same exposed corners over long-times to clean their fur from cave clay or insects. Of importance is the position of the polished areas, which are all only in the cave bear den area of the Saalian ponor level about one meter from the last cave bear walking surface. At several parts of the cave at this elevation such polish seems to have been destroyed or covered by fallen

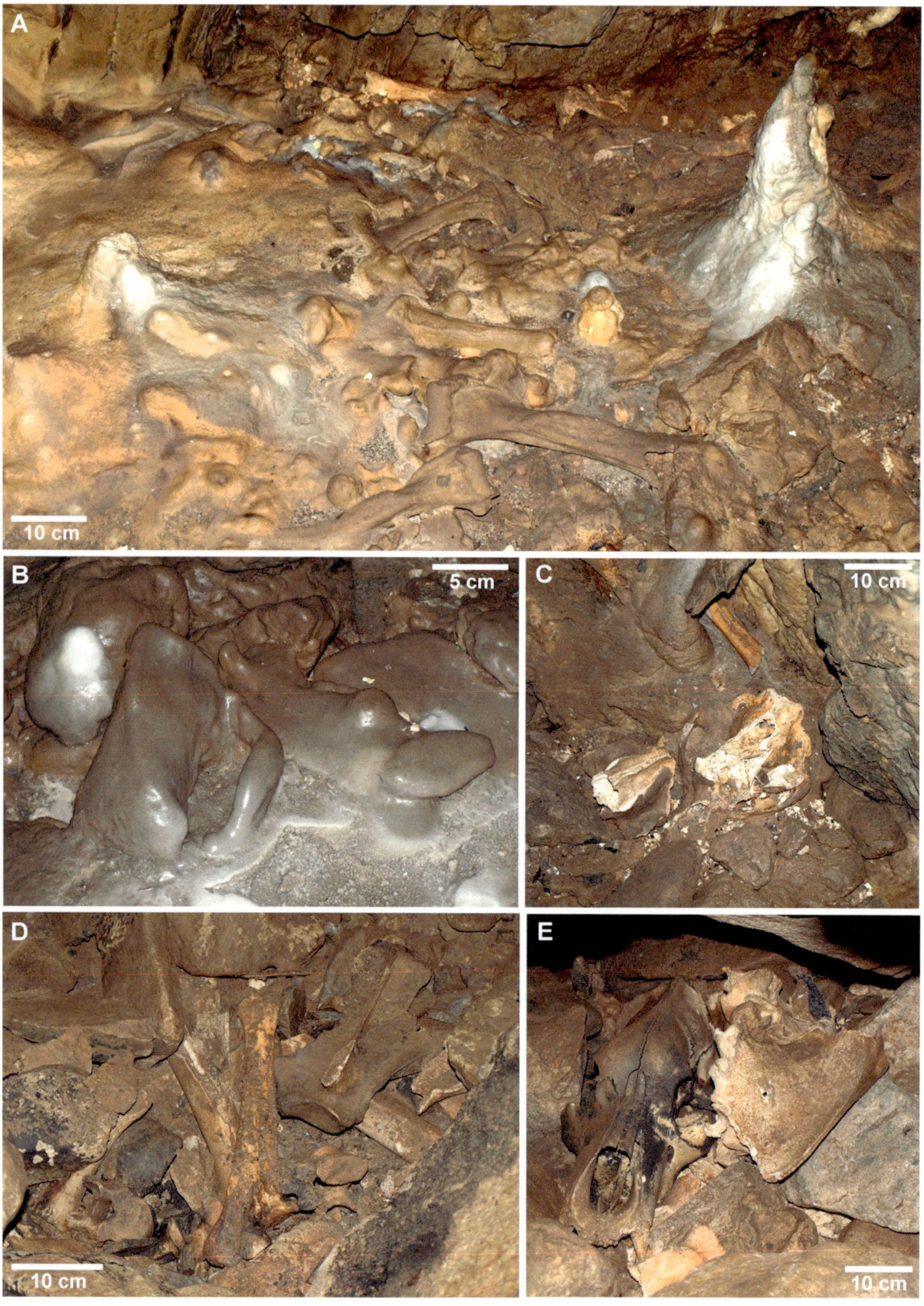
A
10 cm
B
5 cm
C
10 cm
D
10 cm
E
10 cm

Figure 5. Cave bear "bonebeds" in the Hermann's Cave. A. Bones in the Bear Hall by LGM speleothem layer. B. Skull of the large cave bear *Ursus spelaeus ingressus* and besides mandible in the Blue Chamber covered by LGM speleothem layer. C. Weathering or trampling damage of bones (skull) on surfaces, which are not covered by the LGM-speleothem layer. D. Fore limb bones of a cub (ulna, radius, scapula) and E. Nearly complete skull of a young "cave bear" (probably *Ursus spelaeus eremus*) and another braincase not being covered by neither speleothems nor sediment or fallen blocks. Those are still preserved hidden in a niche between the bear and Saal Hall.

blocks/breccias during the LGM times, when ceiling and wall parts collapsed (see Chapters Geology and LGM). Further polished areas are expected to be preserved in some other cave areas of the Hermann's Cave, which have one to two meters sediment refill.

Hibernation Nests

Cave bear hibernation nests are named by the ichnological nomenclature *Ursilorichnus carpathicus* based on more than 140 cave bear nest finds within the Romanian Urşilor Cave system [8]. In the Hermann's Cave, such nests were present at higher numbers, but those were covered by the last speleothem layer, by the LGM cave ceiling collapse large blocks or frost breccia gravel material, or were simply destroyed by caving and different former excavations which focused only on the bones (Kloos, Völker excavations) or archaeological objects (Steiner excavations). Future systematic surface excavations would expose such nests, and in several cases also skeletons or parts within those which are well known for other European caves [8]. Recently, there are only few good examples of hibernation nests in the Hermann's Cave bear den area. One skull and mandible of the large cave bear *Ursus spelaeus ingressus* are in-situ in the Blue Chamber (Fig. **5B**), which were covered by the youngest speleothem layer. Both are within a round hibernation nest area (80 x 80 cm), which nest area was also covered massively by LGM speleothems. Important is its position as it is the last one at the end of the cave 90 m far from the former entrance (Figs. **4**, **8**), where such nests could have been spared from destruction and survived better. As it is known from the Urşilor Cave, each individual was digging the nest by scratching the clay with their short and wide-clawed paws [8].

From this cave mother bears with cubs were mapped in nests lying besides each others [8]. The amount of nests in cave bear dens is an important piece of data in understanding the numbers of hibernating cave bears per winter season, which is not solved yet. As known for northern American black bears which also hibernate in caves similar to cave bears even in deep and far distances from the entrances, those territorial bears are known to use one cave at a maximum with one family - which can consist of maximum five bears, two adults and three cubs [8]. To reach the large cave bear bone amount of older cub, subadult to elderly individuals in

Figure 6. Large cave bear *Ursus spelaeus ingressus* individual skeleton covered by the LGM speleothem, and with hyena scavenging signs on one humerus in the Bear Hall of the Hermann's Cave.

the Hermann's Cave, it would be enough, if only one cave bear died per season in the cave. In some cases, elderly cave bear skeletons are found in their last hibernation nest, whereas there are also examples, that those nests were reused by other bears, which either removed all bones or even slept on parts of the skeletons, mostly the pelvic and hind limb parts, which are sometimes found in the recycled nests in articulation [8].

Three Cave Bear Subspecies

Combining all known fossil cave bear data from DNA, tooth and postcranial bone size metrics and absolute C14-dating [9 - 17], the teeth (especially P4) and skulls from the Hermann's Cave (Tables **1**, **2**) can be identified to their species and subspecies. Their skulls and teeth morphotypes date indirectly the sediments and sections (Fig. **9**). The 16 skulls and 60 new collected P4 teeth allow a more clear identification of those species and subspecies with obvious skull morphology and size characters (Figs. **9-13**). The three Late Pleistocene cave bear species/ subspecies *Ursus spelaeus eremus, Ursus spelaeus spelaeus* and *Ursus spelaeus ingressus* are combined herein in a "cave bear clock" [3] (Fig. **9**), which are compared to absolute dated and DNA-studied European cave bear material [cf. 9 - 17]. Furthermore, the small cave bear *Ursus spelaeus eremus* composite skeleton and other bone material from the Sophie's Cave cave bear dens [4] was used to compare to identify the longbones.

P4-tooth Morphotypes

The cave bear teeth, especially the P4, are used for a relative chronology as result of quite rapid evolution of cave bears at the end of the Middle to Late Pleistocene, MIS 6 - 3 [11]. The P4-index (length to width ratio) is important in order to identify species [9 - 17], whereas there is still a good sampling statistics missing concerning complete skulls including P4 teeth. Generally, only isolated teeth were

analyzed in the past which correlation to skull shapes, remains unclear. The tooth and skull morphotypes are combined for the Herman's Cave material (Fig. **9**) to get a better picture about the species and subspecies cranial osteo-dental morphology of the Hermann's Cave cranial material. Only four- to six-coned tooth morphotypes (= type E after [11]) are found in the Hermann's Cave material, which date into the early to middle Late Pleistocene (Weichselian/Würmian, MIS 5-3) [9 - 17] (Fig. **9**).

<u>Morphotype A-B</u> - The P4 teeth are single-coned, and P4 three main-coned. Those are the older Middle Pleistocene types which resemble to *Ursus arctos deningeri* cave bears and early *U. s. eremus* or *U. s. ladinicus* forms. The late Middle Pleistocene cave bears (Saalian, MIS 8-6) are still unresolved in this problem.

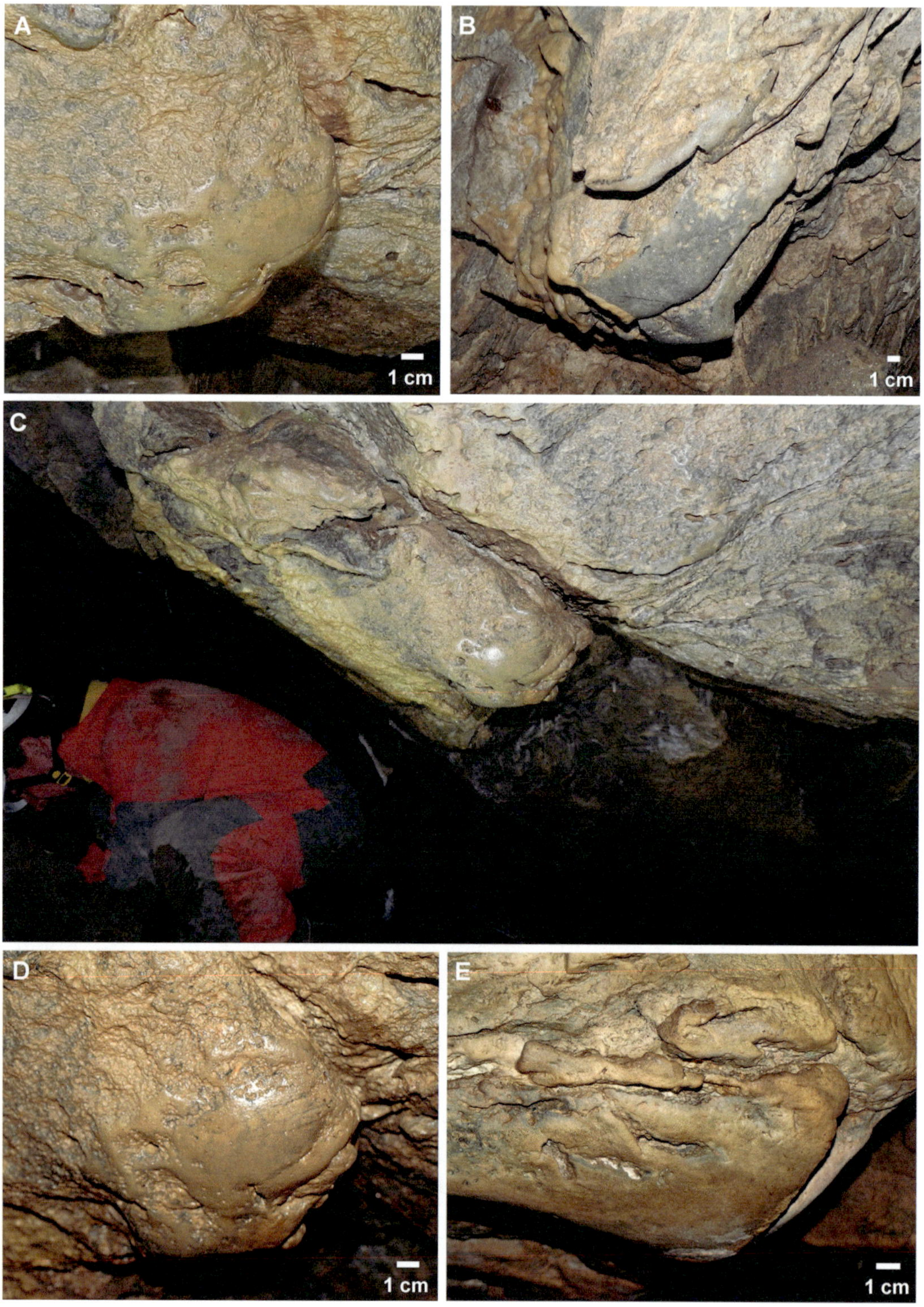
A
1 cm
B
1 cm
C
D
1 cm
E
1 cm

Figure 7. A-E. Examples of cave bear polish in the Bear Hall (see positions in cave map Fig. **5**). Those were produced by fur scratching and on some knob-like corners (here in those cases), or by chance in narrow passages (Schluff Passage).

Figure 8. Example of cave bear hibernation nest in the Blue Chamber at the end of the cave bear den area of the Hermann's Cave (see position in cave map Fig. (**4**). Possibly a complete skeleton is covered by LGM ceiling blocks and breccia and last speleothem layer with candle stalagmites, whereas at minimum the skull is partly visible.

Morphotype C-D - Four-coned P4 are defined to belong to the Late Pleistocene morphotypes C-D and are either from *U. s. spelaeus* or *U. s. eremus* [11] (Figs. **9-13**).

Morphotype E - Five to six multiple-coned P4 upper and lower jaw teeth of the morphotype E [11] are typical for the youngest *U. spelaeus ingressus* species [13]. Such material is found with isolated teeth in the Bear Hall or the Sall Hall (Figs. **9**, **12-15**). The morphotype E is absolute dated around 32-25.000 BP with some examples from skulls in southern German caves [13].

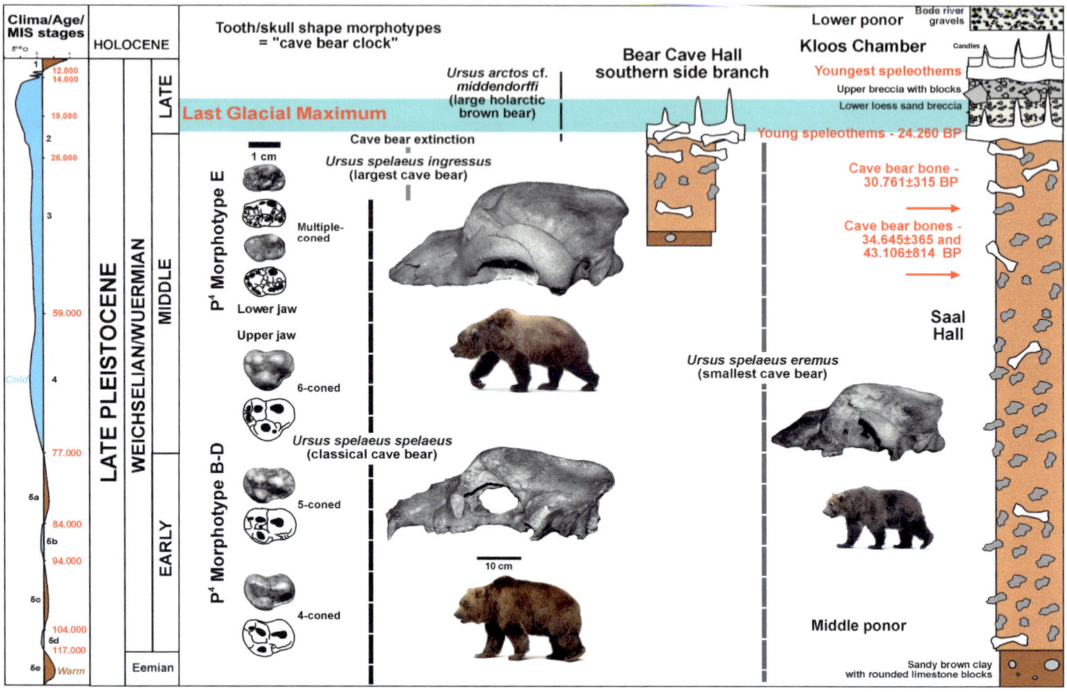

Figure 9. Generalized stratigraphy of the middle ponor level cave bear den and absolute dates on speleothems of the Hermann's Cave. Cave bear species/subspecies skull morphotypes and indirect dating by P4 morphology = "cave bear clock" (composed and modified from [7]).

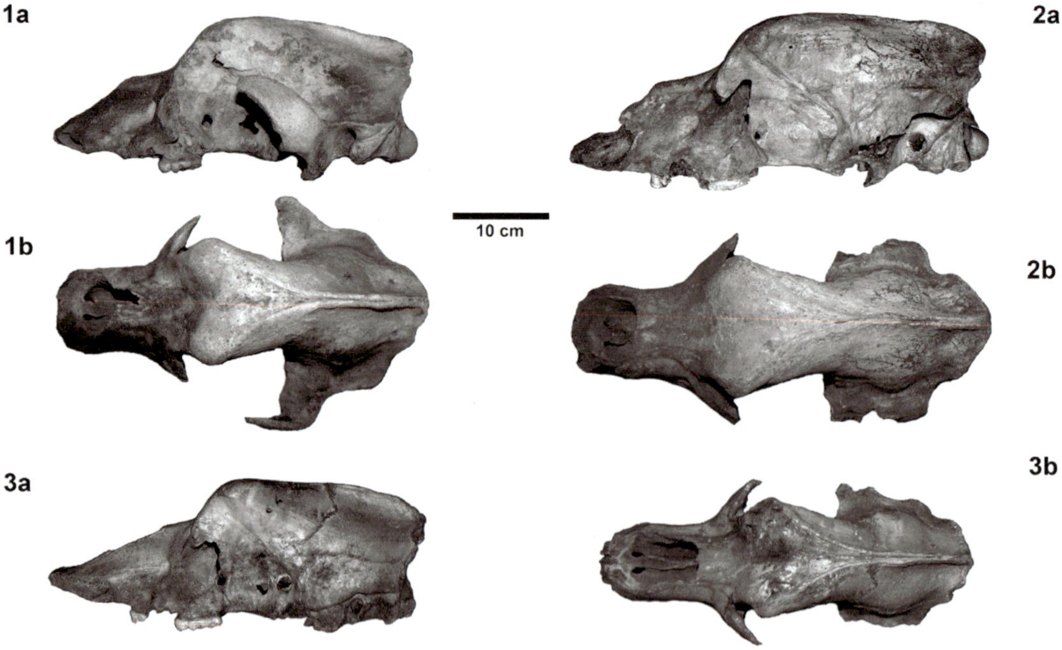

Figure 10. Skulls from the small cave bear subspecies *Ursus spelaeus eremus* fom the Hermann's Cave. 1. Skull from the Saal Hall (coll. LMNB). 2. Skull from the Saal Hall (coll. LMNB). C. Skull from the Bear Hall (coll. RH). a. Lateral, B. Dorsal.

Figure 11. The small speleoid cave bear *Ursus spelaeus eremus* Rabeder *et al.* 2004, with flattened skull and back sloping profile (Illustration R. Uchytel).

Skull Morphotypes

Several of the P4 teeth are preserved in the skull maxillaries which are important to identify the crania to species or subspecies. The skull material presented herein from various collections comprises 16 figured specimens who are more or less complete which can be attributed to three different cranium morphotypes (Figs. **9**, **12-15**). Those three types are also represented in other German Late Pleistocene cave bear dens such as *e.g.* the Sophie's Cave [4, see **Vol. 1** of this book series], or the Zoolithen Cave [18].

Morphotype A - The plesiomoph and older type dates back to the smaller Middle Pleistocene *U. arctos deningeri*, which is not represented in the Hermann's Cave. The skull shape is in lateral view generally dorsal convex. The only well preserved German skull that was compared from the Zoolithen Cave [18] has the small P2-3 in its maxillaries, whereas an unpublished mandible from the Zahnloch Cave has also those premolar teeth. The skulls are even smaller or similar in length as the advanced spelaeoid small cave bear *U. s. eremus* skulls.

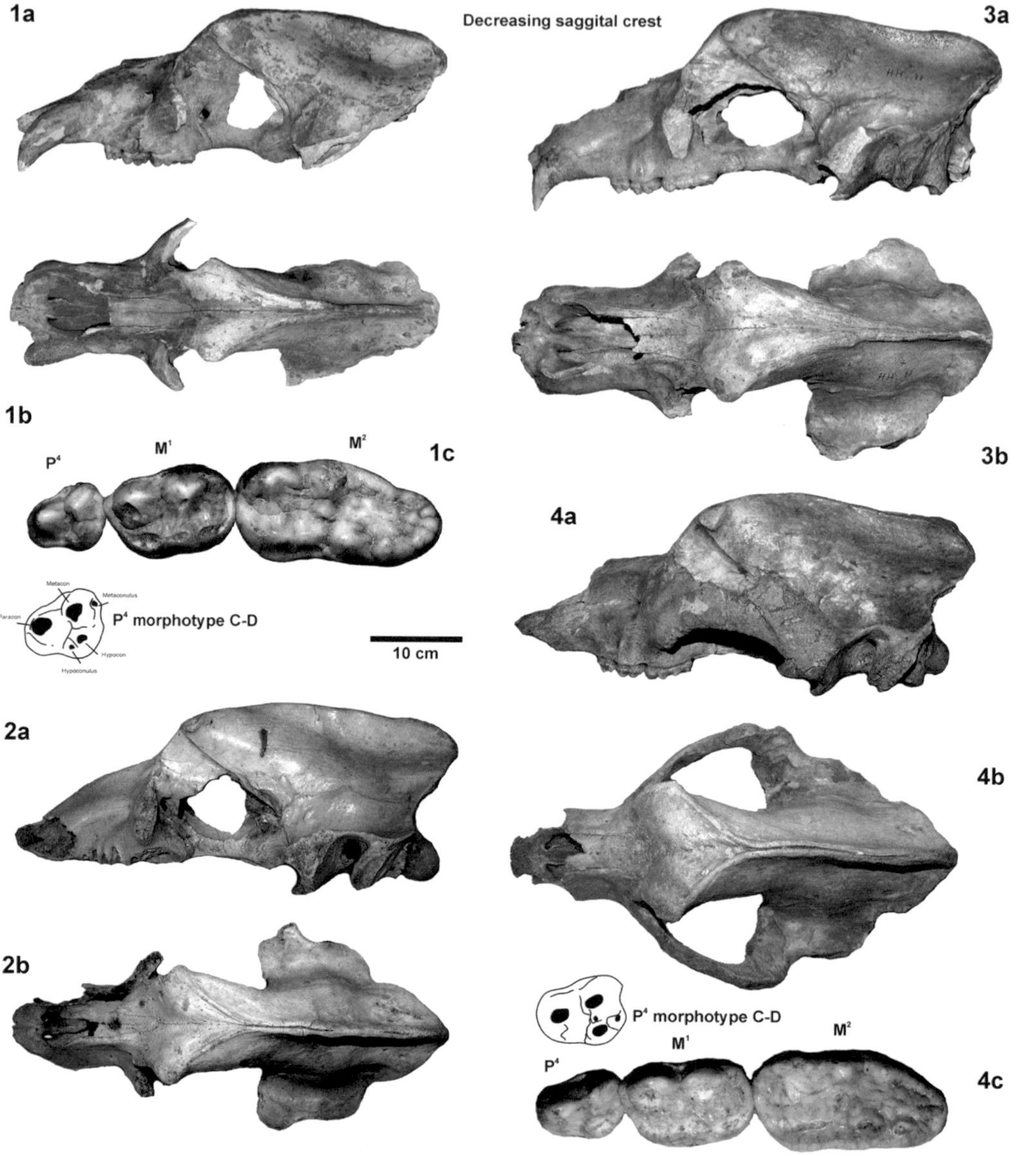

Figure 12. The medium-sized *Ursus spelaeus spelaeus* with four to five coned P4 ad steep decreasing sagittal crest from the Hermann's Cave. Skull of an adult cave bear (coll. MQ). 1. Skull of an adult cave bear (coll. MQ). 2. Skull of a subadult from the Saal Hall (coll. NMB). 3. Skull of an adult cave bear (coll. LMNB). 4. Skull of an older adult from the Bear Hall (bone dump) (coll. RH). a. Lateral, b. Dorsal, c. Occlusal.

Figure 13. The classical European cave bear *Ursus spelaeus spelaeus* (Illustration R. Uchytel).

Morphotype B - Smallest spelaeoid cave bear skulls in total length measure in average between 370 - 390 mm and are from *Ursus spelaeus eremus* (Figs. **10**, **11**). It is the smallest European cave bear form which is known by a composite skeleton of the Sophie's Cave [4]. This subspecies occurred in middle high elevated European mountain regions of maximum 1.100 m in northern Germany, and a few higher in southern Germany distributed cave bear. Their P4 teeth represent the C-D morphotypes, whereas in some cases the P3 can be present as a rudimentary relict tooth [11]. Those small bears might have evolved from the small Middle Pleistocene *Ursus arctos deningeri* species [15 - 17]. The skulls of *U. s. eremus* are quite different to those with having the typical spelaeoid steep frontal (Figs. **12-15**) and reduced anterior P2-3 teeth. It seems, the evolution between both happened within the Middle Pleistocene between the Elsterian to Saalian (Scandinavian glacier terminology glacial periods sensu [15 - 17]), which is a remaining question that cannot be solved by DNA. Those cave bears seem to represent the small cave bear forms which lived sympatric to the high alpine small cave bear subspecies *Ursus spelaeus ladinicus* [15].

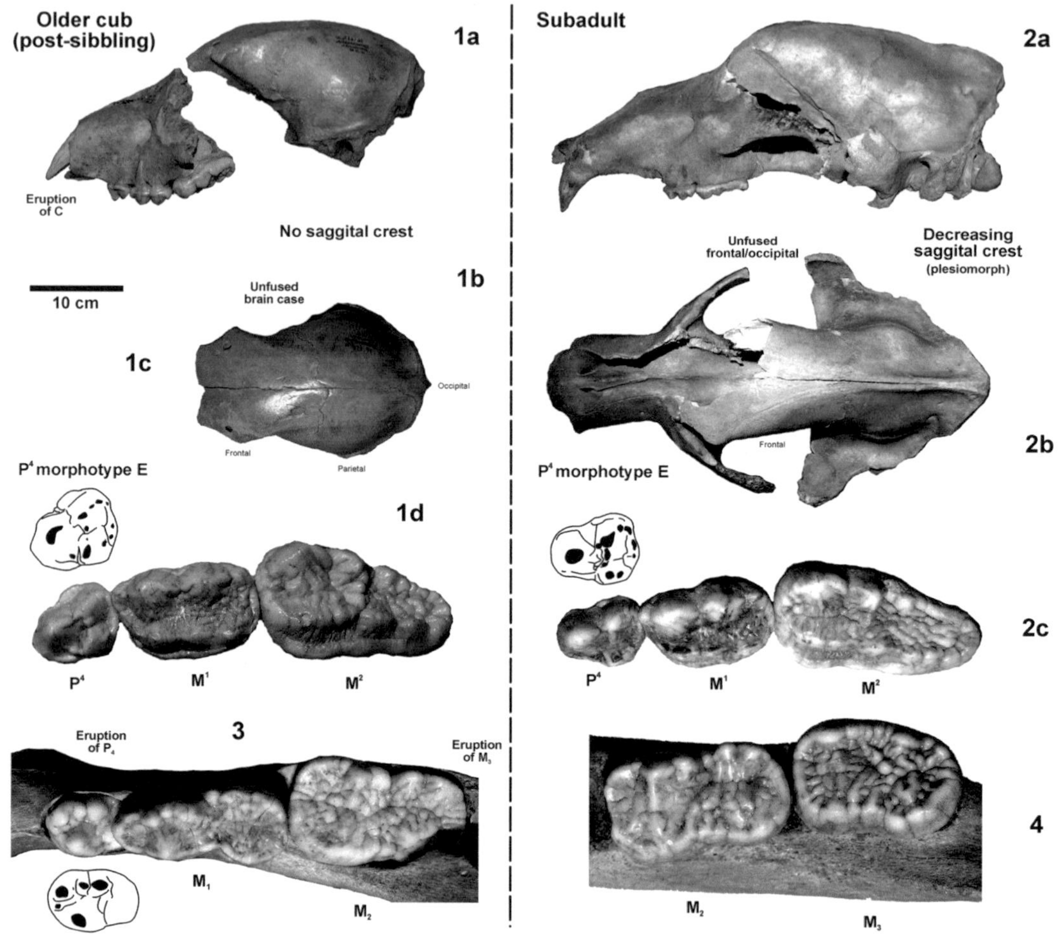

Figure 14. Largest cave bear *Ursus spelaeus ingressus* skulls from the Hermann's Cave with multiple-coned P4. The skull shape changes within the ontogeny from the spelaeoid shape with still posterior decreasing crest to horizontal shapes in adults to elderly. 1. Composed skull consisting of a brain case and a maxillary, both from the Saal Hall (coll. LDA).. 2. Skull of an older cub-subadult from the Bear Hall (coll. RH). a. Lateral, b. Dorsal, c. P-M teeth occlusal. 3. Left mandible of an older cub with final lower jaw tooth eruptions of the P4 and M3 from the Saal Hall (coll. LDA). 4. Left mandible M2/3 of a subadult with full changed teeth from the Saal Hall (coll. LDA). a. Lateral, B. Dorsal, c. Occlusal.

Morphotype C - Middle-sized skulls are in average in total length between 420-470 mm with strong decreasing sagittal crest are from *Ursus spelaeus spelaeus* (Figs. **9**, **11**, **13**). In dorsal view, the triangular shape and length of the frontals and length of the sagittal crest vary strongly (Fig. **12**). The morphotype of long-triangular frontals reaching over the middle of the sagittal area (Fig. **12**) is rarer, but was also found in the mounted unpublished composite skeleton being exposed in the Grosse Teufels Cave (Bavaria).

Morphotype D – The first herein figured ontogenetic stage of *Ursus spelaeus ingressus* is from post-sibling cubs (about 3 months), which are the final tooth change with the canines, whereas the premolar and molar maxillary teeth are full changed and still unused (Fig. **14.1**). The skull of an older cub to subadult (about 12 months) is in the spelaeoid stage (repeating evolution) demonstrating is evolutionary relationship to *U. s. spelaeus* (Fig. **14.2**). The large skulls of adults may measure in average length around 470-520 mm (Figs. **15**, **16**). The largest skull ever measured from a German cave is a male skull from the Hermann's Cave that reach giant size of 540 mm length (Fig. **16.1**). The crania have in lateral view a nearly horizontal sagittal crest and a step-like frontal (angle around 70°). Their P4 morphotype E [11, 13, 17] has multiple-coned and enlarged cusps as an adaptation to full herbivory [13]. The material from the Hermann's Cave can be compared to a figured and 32-24.000 BP absolute dated skull from a southwestern German cave bear den [13]. Those youngest in Europe appearing or developing cave bear species seem to originate from *U. s. spelaeus* [11, 13, 17].

Postcranial Cave Bear Size Differences

Postcranial complete longbone material from the former bone dump in the Bear Hall such as the material from the Steiner and Völker excavations was used for size estimation correlation to the three skull-based species. Selected longbones of each size class were figured and compared osteometric (Fig. **18**, Table **1**). Further metric studies have to be made on smaller pedal bones (*e.g.* metapods), which will be only important for future composite skeletons for exhibition.

Small - The smallest longbones fit to the smallest skulls, which are described herein as *U. s. eremus*. The material is similar in size ranges to the figured skeleton and bones from the Sophie's Cave [4] (Fig. **18**). Similar small sized humeri (Fig. **18K**), ulna (Fig. **18L**), radius (Fig. **18**), femur (Fig. **18**) and tibia (Fig. **18**) bones can be reported herein from the German Perick Caves, Keppler Cave, the Zoolithen Cave, Grosse Teufels Cave, or Hoehling Cave.

Medium-large sized – The postcranial longbones as figured with the humerus, ulna, radius, femur and tibia are much larger and closer in proportions to the largest forms. From the size those correlate in proportions to the *U. spelaeus spelaeus* skulls. Such proportions are found in the Perick Caves or the Zoolithen Cave.

Large – Most large cave bear longbones are common in the Hermann's Cave and are figured again with all five large longbones (Fig. **18**). In some cases it is and will become difficult to distinguish largest *U. s. spelaeus* male bones from small *U. s. ingressus* females.

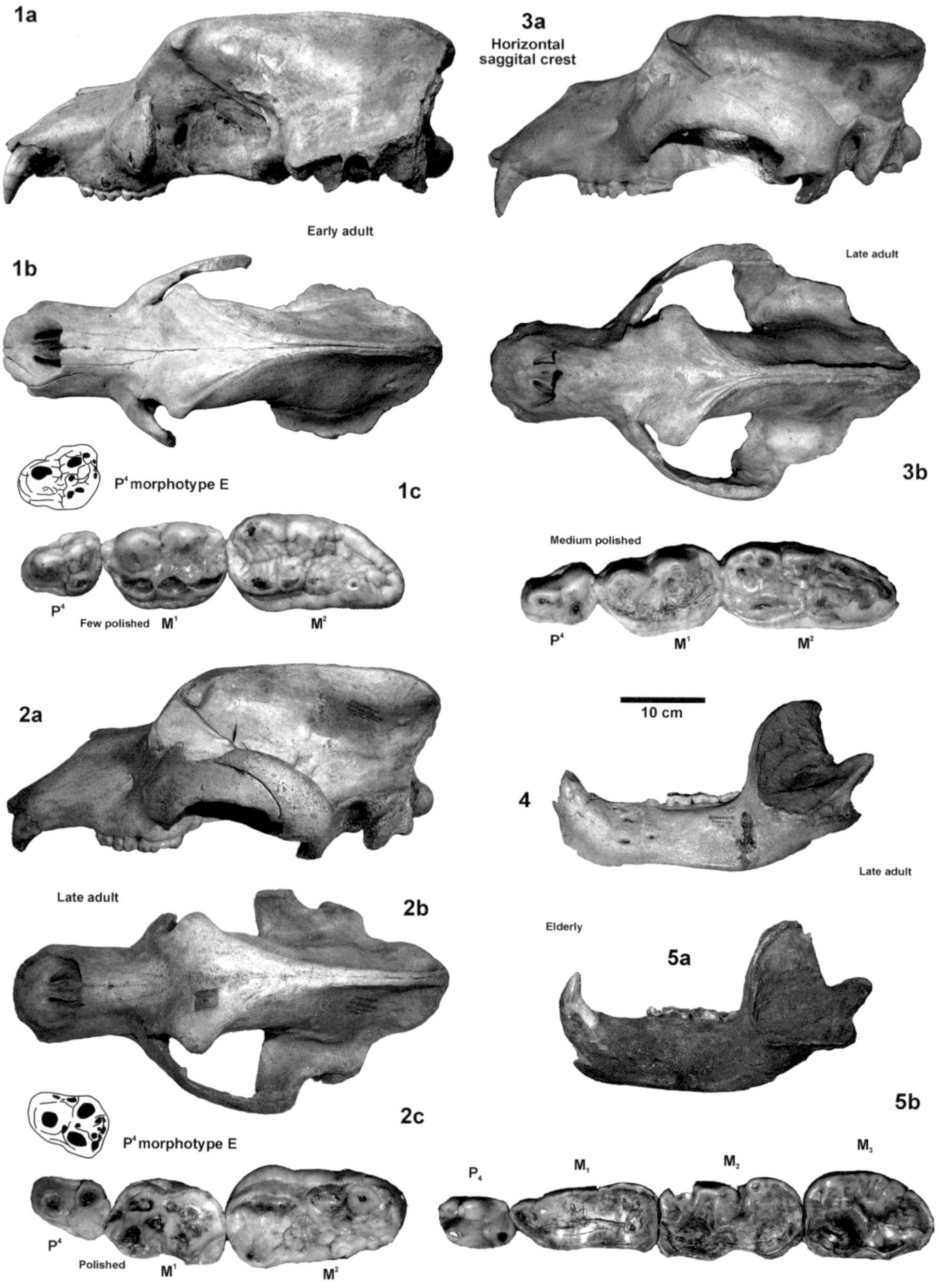
1a
3a
Horizontal
saggital crest
Early adult
1b
Late adult
3b
P4 morphotype E
1c
Medium polished
P4
Few polished
M1
M2
P4
M1
M2
2a
10 cm
4
Late adult
Late adult
2b
Elderly
5a
5b
2c
P4 morphotype E
M3
P4
M1
M2
P4
Polished
M1
M2

Figure 15. Largest cave bear *Ursus spelaeus ingressus* skulls from the Hermann's Cave with multiple-coned P4 and horizontal sagittal crest. The skull shape changes within the ontogeny from the spelaeoid shape with still posterior decreasing crest to horizontal shapes in adults to elderly. 1. Skull of a subadult bear from the Saal Hall (coll. RH). 2. Skull of a late adult bear from the Bear Hall (coll. LMNB). 4. 2. Skull of a late adult bear from the Bear Hall (coll. LDA). 4. Left mandible of an elderly from the Saal Hall (coll. LDA). 5. Left mandible of an elderly from the Bear Hall (coll. LDA). a. Lateral, B. dorsal, c. Occlusal.

Neonate Aborts or Siblings

Isolated neonates, siblings and very young cave bear cub bones or deciduous teeth are present all over the cave bear bonebed sediments of the Hermann's Cave. Those are found after sieving sediments, which was not made in historical or past excavations [2, 5, 6]. First sieving tests were made in 2011 with fraction above 1 mm on 10 liters of cave bear bonebed sediment from the Bear Hall. Several smallest cave bear bones and some deciduous teeth were isolated [7]. The herein figured cave bear skeletons from the Saal Hall are in different age classes of a neonate unborn or stillborn or aborted (Fig. **19A**) and sibling with deciduous dentition (one preserved dm4, Fig. **19B**). They were composed from material of the Steiner excavation which was found 90 meters away from the former cave bear den entrance [19]. Such incomplete neonate to very young cub individual skeletons have been figured and reported as “cave bear babies“ in Europe only from two other German caves, the Dechen Cave [20] and the Peters Cave [21]. Indeed, those represent in many cases aborted or stillborn. Therefore such bones and deciduous teeth are figured and described from many European cave bear

Table 1. Cave bear leg longbone metrics to bones of Fig. (17) (Length/distal width in mm).

Subspecies	**Bone Type**				
	Humerus	**Radius**	**Ulna**	**Femur**	**Tibia**
U. s. ingressus	510/140	345/90	390/65	500/130	330/90
U. s. splaeus	470/150	310/80	345/58	460/120	280/80
U. s. eremus	360/113	290/70	330/45	360/70	265/70

Table. 2. Cave bear bones from the autochthonous Late Pleistocene (MIS 3-5e) cave bear den bonebed layers of the Hermann’s Cave, Rübeland, Harz Mountain Range (northern Germany).

Species	**Amount/Cave Area**	**Collection**
Ursus spelaeus spelaeus/eremus/ingressus	1.250 (excavation Steiner 1961), 2.230 (excavation Völker 1984/85), 1.430 (bone dump 2012) = 4.910 (total)	LMNB, LDA, RH, MB and in cave

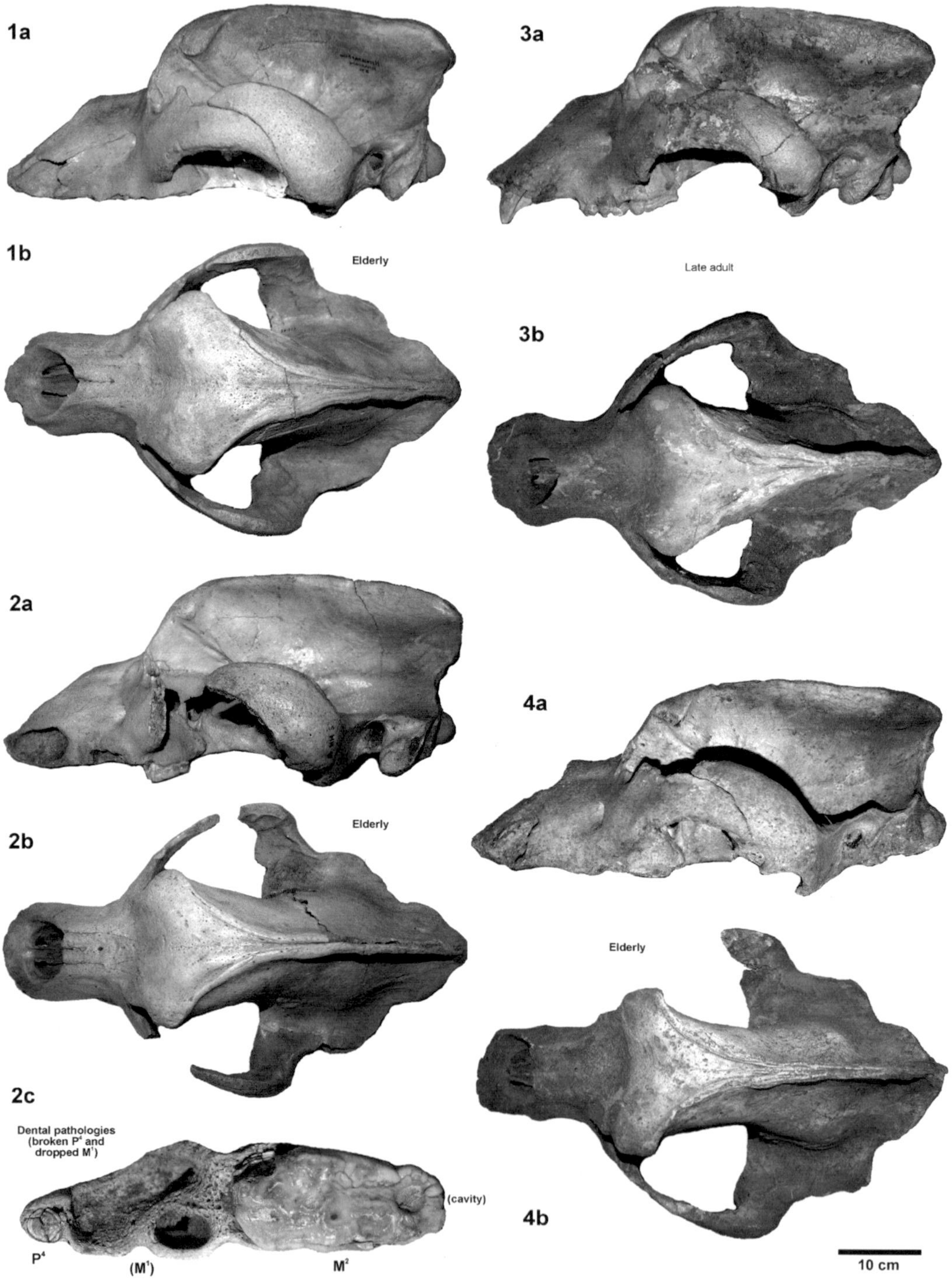
1a
3a
1b
Elderly
Late adult
3b
2a
4a
2b
Elderly
Elderly
2c
Dental pathologies
(broken P4 and
dropped M1)
(cavity)
4b
P4
(M1)
M2
10 cm

Figure 16. Largest cave bear *Ursus spelaeus ingressus* skulls from the Hermann's Cave with multiple-coned P4 and horizontal sagittal crest. The skull shape changes within the ontogeny from the spelaeoid shape with still posterior decreasing crest to horizontal shapes in adults to elderly. A. Skull from the Saal Hall (coll. LDA). B. Skull of an elderly bear (coll. LMNB). C. Skull of a late adult bear from the Bear Hall (from skeleton of the composed skeleton, Figs. **3C, D**, coll. RH). D. Skull of late adult bear (coll. LMNB) (skulls partly mirrored). a. Lateral, B. Dorsal, c. Occlusal.

Figure 17. The last European and largest herbivore cave bear *Ursus spelaeus ingressus* with its typical stepped frontal and high shoulder (llustration R. Uchytel).

dens [22]. Stillborn births must have happened quite often in the cave, especially when the mother bear was not strong enough after the long hibernation time. The lack of detail studies of such neonate-sibling material is a problem for the exact aging of the individuals, especially their postcranial bones, and the palaeoecological interpretation or statistics about real non-carnivore impacted real natural death birth rates. The neonate and sibling skeletons from the Saal Hall improve the knowledge about the birth and raising areas within larger caves. The birth must have happened at the hibernation pits and even at deeper and farther distances from cave entrances. The first weeks or months cave bears might have stayed in those deeper cave parts, which were the more secure place to protect themselves against different predators [23]. The two skeletons, the larger amount of such neonate-sibling aged skeleton and bone remains from other caves and large amount of deciduous teeth from European cave bear dens allow speculating, that cave bear cubs did not leave the cave within their sibling period [20, 23], and possibly even not the first three months.

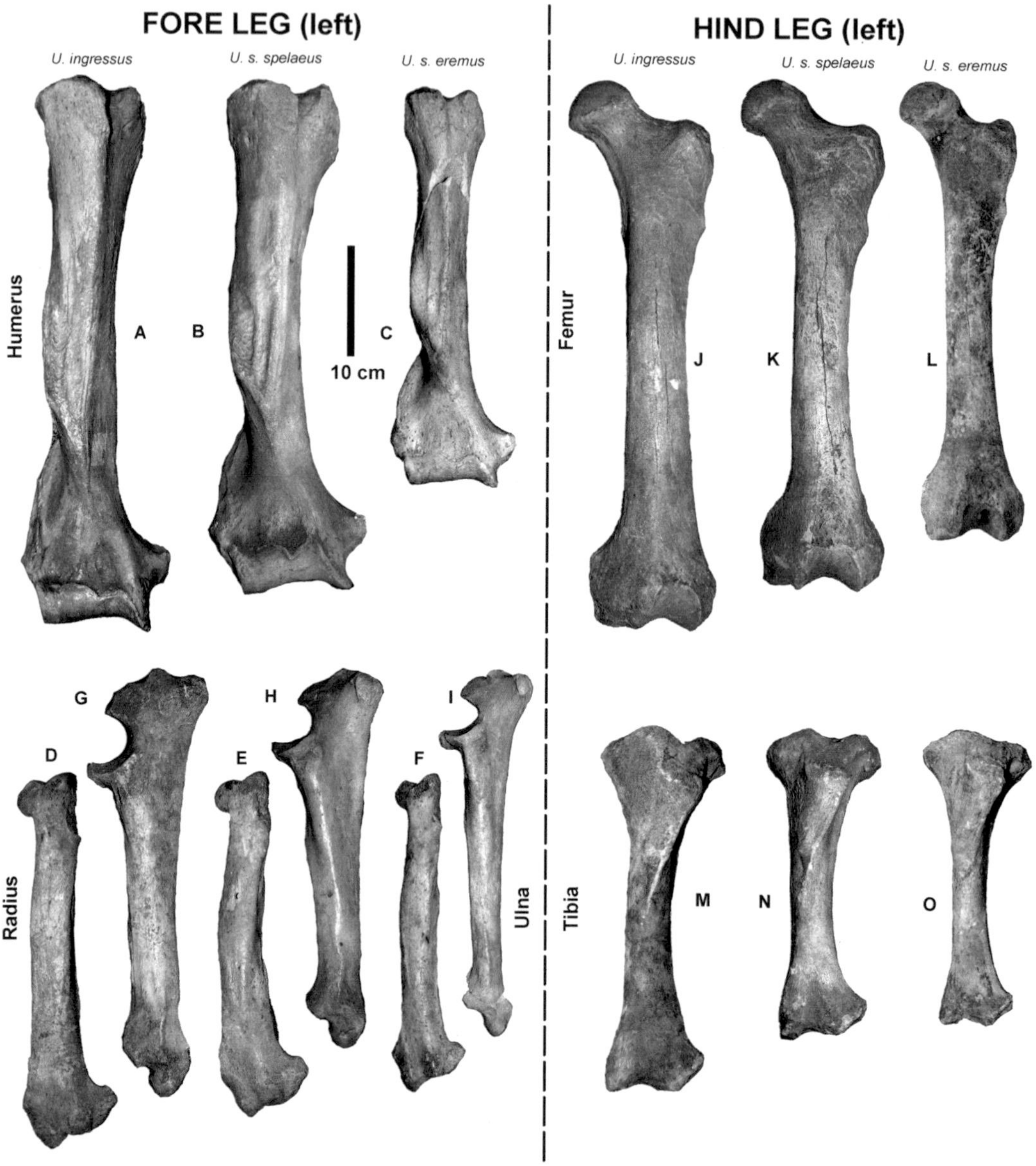

Figure 18. Comparison of the postcranial longbone cave bear species and subspecies proportions between the largest *U. s. ingressus*, medium-large *U. s. spelaeus* and small *U. s. eremus* (cf. metrics in Table **1**). All material from the bone dump of the Bear Hall, coll. RH).

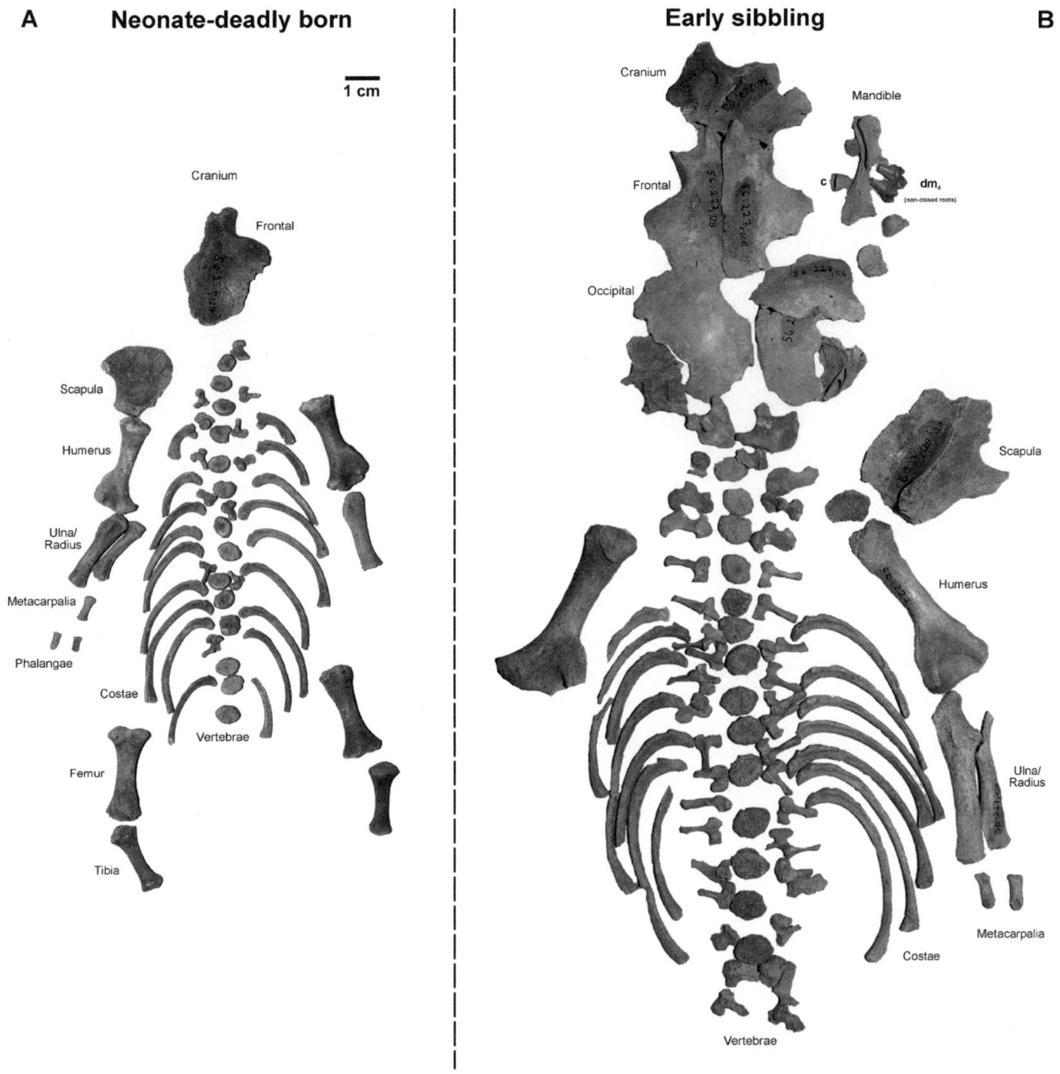

Figure 19. Two different arranged aged incomplete cave bear neonate/cub skeletons from the Saal Hall, Hermann's Cave (coll. LDA). A. Neonate or aborted stillborn. B. Sibling in deciduous teeth dentition most probably *U. s. ingressus* with preserved lower jaw dm4 (non-closed roots).

Cave Bear Cub Mortalities

Mortality analyses of cave bears were often used incorrectly for palaeoanthropological interpretations by archaeologists [32, or by palaeontologists [33 - 35], because they did not incorporate the top predators such as *Panthera leo spelaea* (Fig. **20**; and chapter Lions,) and scavengers: Ice Age spotted hyenas and Ice Age wolves. *Crocuta crocuta spelaea* and *Canis lupus spelaeus* as main cave

Figure 20. Ice Age steppe lion on its cave bear kill. He consumes first the intestines and inner organs, and possibly some meat using its meat-cutting dentition ("digital cave imaging" illustration by R. Uchytel).

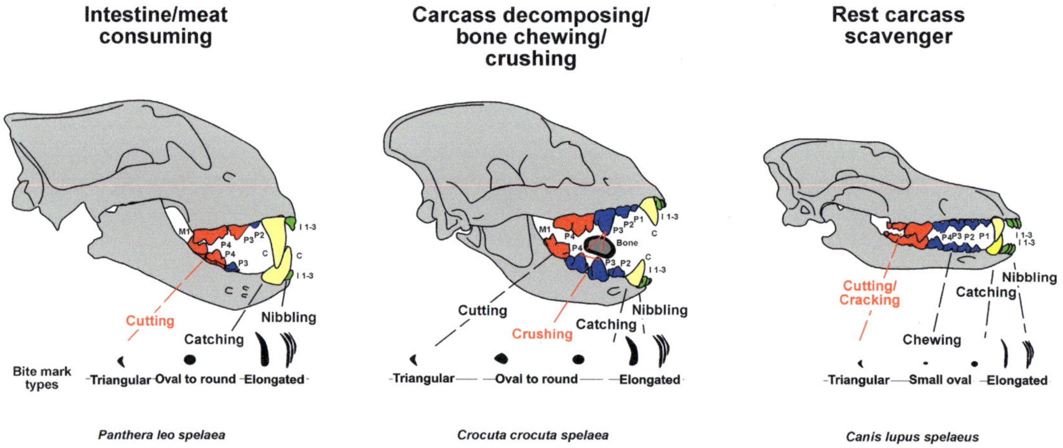

Figure 21. Differences in bite and jaw/tooth specializations (dentitions) of the three large Ice Age predators, (modified from [4]).

bear killers and cave bear scavengers [24] (Fig. **21**). Mortalities of the different cave bear species of the Hermann's Cave are not solely signs of carnivore origin. The few secure human caused mortalities (or/and butchering only) are shown on a few single bones (see chapter Aurignacien cave bear hunters). The proof for a cave bear kill by a top predator is anyway impossible to present, because bite marks cannot be distinguished in their pre- or postmortal origin. This makes it impossible to attribute in many cases single bite marks or bone damaged bones exactly to the two predators, hyena and wolf (cf. similar shaped bite marks in Fig. **22**). In the cases of round-oval bite holes which are found commonly in cave bear cub longbones scavengers, especially the femorae (Figs. **22H**, **23A, B**), there is only one candidate, who could have produced such holes – hyena scavengers [24].

Top Predators - Cave Bear Killers and Scavengers

First large carnivore bite marks on cave bear bones from the Hermann's Cave were figured in a few examples by former excavators [6, 19]. The predators were then identified on the abundant incomplete cave bear bones from all excavation and the Bear Hall bone dump material in more recent studies [7]. Herein figured are selected bite damaged crania and lower jaws such as the longbone damage stage A-C series (humerus, ulna, radius, femur and tibia) such as axial skeleton elements (vertebrae and pelvis) of different aged individuals (Fig. **22**).

Crania and Mandibles

Most of the cub crania or bone elements expose bite scratches, holes or other bite marks. The herein figured cub skull has an oval 6 x 10 mm puncture hole in the frontal (Fig. **22A**). Also the jugals have bite damages. Such puncture holes of cave bear cub crania are recently figured for other caves as well and seem to have resulted from the predators canine teeth [23, 24].

Those can even result of killing bites by steppe lions or Ice Age spotted hyenas, which canines fit in diameter sizes to such bite holes [24]. In some cases such holes can also result from the premolar crushing triangle teeth of hyenas, if those crushed the skulls and especially brain cases to reach the brain marrow [23, 24]. To such cranial scavenging signs on cubs also puncture holes of similar large sizes are found as figured herein for one sibling (Fig. **22D**) and one cub mandible (Fig. **22C**). The mandible from the Saal Hall of the sibling with two large carnivore impact holes at the ramus was damaged most probably by the two hyena premolar crushing teeth P3-4. Also the mandible fragment from the Lower Fluvial Passage cone with two large carnivore impact holes was again most probably punctured by the hyena premolar crushing teeth P3-4. Similar sized and positioned puncture holes on cave bear cub crania are of well documented and identified as hyena origin are figured for other larger cave bear dens of Germany

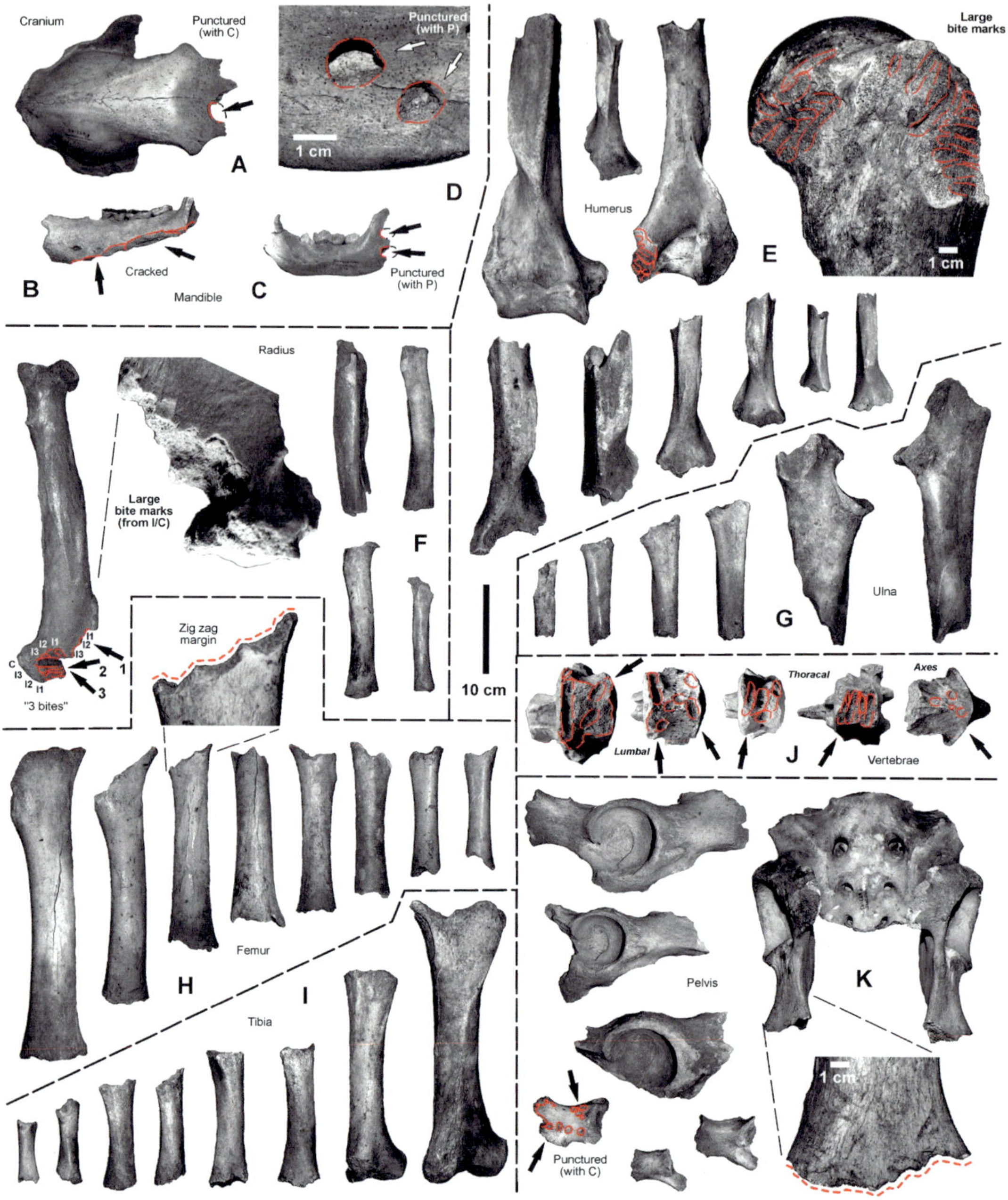

Figure 22. Bite and chew damaged cave bear bones of "cave bear" cubs and adult individuals from the Hermann's Cave. A. Skull of a cub (about 3 months in age) with canine bite impact of a hyena or lion from the Saal Hall (coll. LDA). B. Crushed mandible of an elderly individual from the Lower Fluvial Passage cone (coll. RH). C. Mandible of a sibling with two large carnivore impact holes (most probably hyena premolar crushing teeth P3-4) from the Saal Hall (coll. LDA). D. Mandible fragment with two large carnivore impact holes (most probably hyena premolar crushing teeth P3-4) from the Lower Fluvial Passage cone (coll. RH). E. Humeri with chewed proximal joints and bone shafts of different aged individuals from different cave

parts (coll. RH). F. Radii with bite marks and shafts of different aged individuals from different cave parts (coll. RH). G. Ulna shafts of different aged individuals from different cave parts (coll. RH). H. Femur shafts of different aged individuals from different cave parts (coll. RH). I. Tibia shafts of different aged individuals from different cave parts (coll. RH). J. Vertebrae with large carnivore bite impact and scratch marks on the ventral sides of different aged individuals from different cave parts (coll. RH). K. Chew and bite damaged pelvis and coxa of different aged individuals from different cave parts (coll. RH).

such as the Perick Caves, the Grosse Teufels Cave, the Zoolithen Cave, or the Sophie's Cave [4]. In adult cave bear skulls such holes are extremely rare. Those more often have chew damaged jugal arches as result of the lower jaw removal by the predators. Mainly hyenas seem to have developed the strategy to remove in similar ways lower jaws from cave bear skulls (and even cannibalistic from themselves or other guilt [22 - 25]), which also result in the ramus damage of the mandibles. The crushing of the mandibles happened behind the M3 tooth, which resulted in diagonal breakage on adult to elderly mandibles (Fig. **22B**) [22 - 25]. In the case figured herein, the mandible was even further crushed in the ventral parts, where zigzag margins indicate its hyena crushing teeth origin [24, 26].

Fore Limb Longbones

Humeri with chewed proximal joints and remaining bone shafts of different aged individuals from different cave sections expose large predator caused typical zig-zag margins or bite mark impressions (Fig. **22H**) [26, 27]. Radii with bite marks and shafts of different aged individuals from different cave parts (Fig. **22J**) are also common. Most interesting is one radius of a larger cave bear species that allow the reconstruction of three bites at its distal joint (each I1-C, Fig. **22F**) of the anterior dentition of a large predator, either lion or hyena. The cave bear ulna with three bites on its distal joint is quite unique in the record, because it allows identification between the anterior C and I1-3 teeth of the predator (Fig. **22F**). Their large sizes and distances exclude wolves [27], but do not allow distinguishing between steppe lion or Ice Age spotted hyena which is better to demonstrate on the largest guilt [28, 29]. One of the carnivores was biting several times with its "snout" into the bear fore leg/manus area. This is quite unusual in the scavenging process. Normally, the predator would have used the breaking scissor dentition to cut of the paw. Ulna shafts of different aged individuals from different cave parts (Fig. **22G**) have similar damage as all other longbones.

Axial Skeleton

Vertebrae with large carnivore bite impacts and scratch marks on the ventral sides of different aged individuals from different cave parts of the Hermann's Cave (Fig. **22J**). The 8-12 mm wide bite puncture holes and scratches are indeed canine caused bite holes, but those were produced only in the soft spongiosa of the vertebrae centra. This is of importance for the understanding of the feeding strategy of the top predators – lions and hyenas, whose bite holes and scratches on the ventral side of vertebrae, which are found mainly in the lumbar to middle thoracic body cavity region (Fig. **22J**), similar as described for other cave bear dens [4, 18, 24, 25, 30]. Similar bite damages on vertebrae of the small cave bear *U. s. eremus* were also figured for the Sophie's Cave [4]. The third larger cave bear carcass feeder was the wolf, which was best documented on the Sophie's Cave, where in wolf faecal pellets and the wolf den many small cave bear bones along with stomach acid corroded surfaces have been found [4]. In addition the Ice Age wolf *Canis lupus spelaeus* has a quite large breaking scissor dentition and was also able to leave only bone shafts of larger longbones (Fig. **21**) [4]. Similar sized predators (lion/hyena) bite punctures and scratches are also found on Late Pleistocene forest elephant and mammoth carcasses vertebral columns on the ventral sides between the thoracic and lumbar vertebrae [28, 29].

Some of the 5 - 8 mm wide bite scratches/holes sometimes can be reconstructed in their distance metrics to the predator size, which correlate to the canine distance (upper or lower jaws) similar in hyenas or lions, especially considering the sexual dimorphism in each predator group: a small female lion is close or similar to large female hyena canine distances [28, 29]. Wolves can be excluded as producing those large sized and wide canine bite metrics in several cases, which was discussed on material of *U. s. eremus* from the Sophie's Cave [4]. The bite scratches are found often in cave bear dens on cave bear vertebra (ventral side) and prove the scavenging of the inner organs/intestines and eating into the body cavity [4]. As suggested by chew and bite damaged pelvis and coxae of different aged individuals from different cave parts (Fig. **22K**). Similar acetabular scavenging signs are known from the wholly rhinoceros, or wholly mammoth which damaged bones have been found at hyena dens [29, 31].

Hind Limb Longbones

Femur and tibia shafts with larger bite marks of different aged individuals were also found all over the cave (Figs. **22.9, 22.10**). At all European cave bear dens those damaged cave bear bones are typically present in larger amounts of 15-25% of all longbones [22 - 25], but have been not collected in former excavations due to their "incompleteness". Such bone shafts with chewed joints are also common in the Sophies's Cave cave bear den [4], especially on cave bear cubs [24], which can show up to 80% damaged as calculated herein on the material from the Hermann's Cave. Identical damaged cub and adult individual bones are figured now from several European caves which result from the breaking scissor teeth of the large predators [22 - 25, 27].

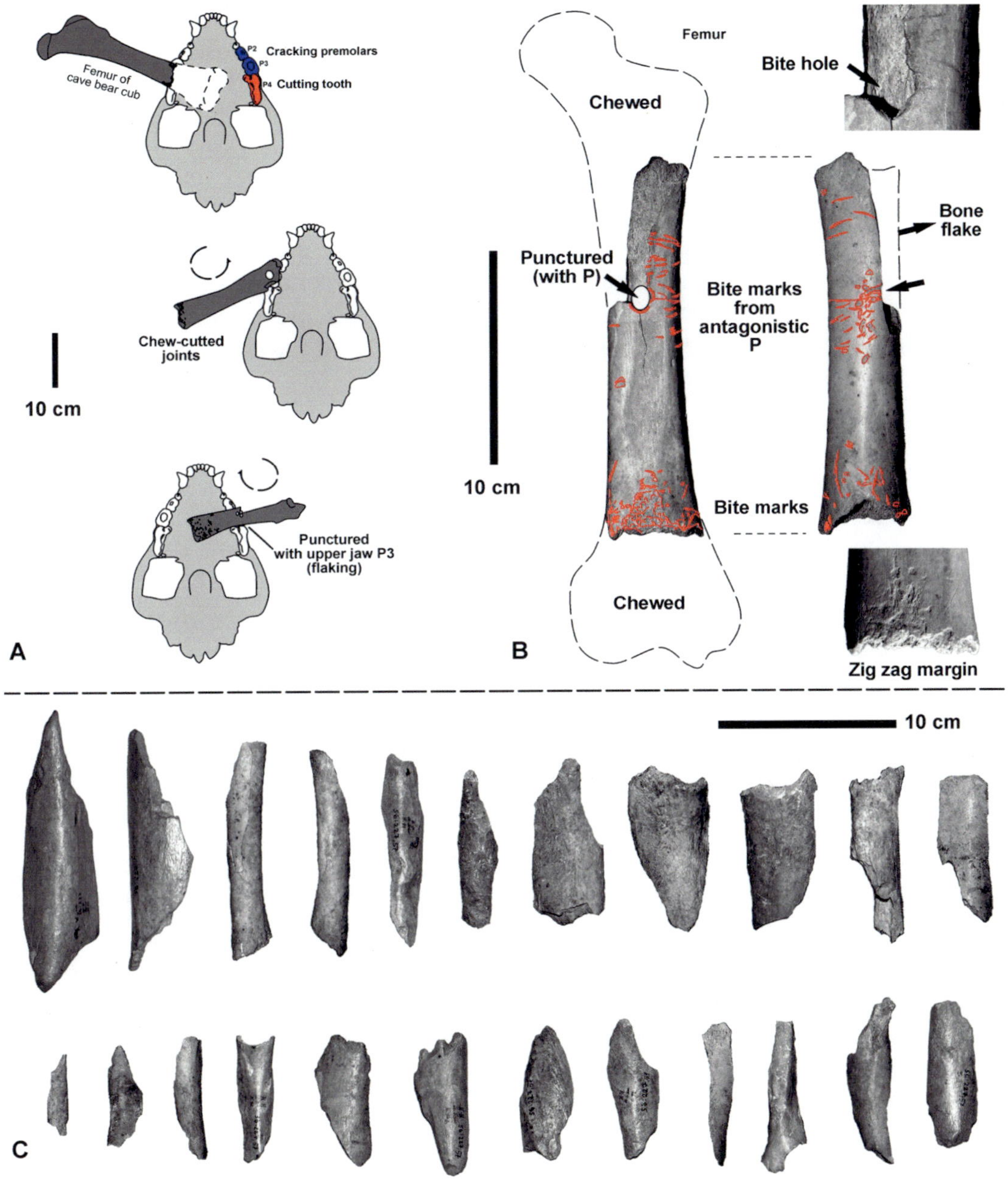

Figure 23. A. Model of non-successful bone crushing of cave bear cub femora by Ice Age spotted hyenas. After chewing the soft spongious distal joints the compact shaft did not crack after puncturing it with the cracking premolar teeth. B. Partly cracked cave bear cub femur shaft from the Saal Hall, Hermann's Cave (coll. RH). Such femora with holes were even misinterpreted as "Neanderthal bone flutes". C. Cave bear longbone compacta fragments (RH, LDA).

The misidentified "Neanderthal – bone flutes" are simply the product of Ice Age spotted hyenas, which used the premolar triangle cracking dentition to crack the longbone shafts while scavenging [24]. As in several publications presented, the conical hyena premolar crushing teeth (Figs. **21**, **22A**), produced such bite holes in shafts. Herein such a "pseudo bone flute" is presented from the Bear Hall where only a part of an older cub femur splinted (Figs. **23A, B**). The hole making premolar antagonist teeth left on the opposite side also bite impact and scratch marks. Such punctured, partly flaked or completely in pieces flaked (Fig. **23C**) cave bear cub femora are in detail discussed for several cave bear site of Europe [22 - 25] which show that at some sites indirectly the presence of hyenas (Fig. **24**), which often are not found with bones more far from the entrance at such cave bear occupied large cave systems [22].

Figure 24. Ice Age spotted hyenas destroy and damage a cave bear carcass. With the crushing dentition they crack longbones ("digital cave imaging" illustration by R. Uchytel).

Cave Bear Pathologies

Some examples of dental and postcranial osteopathologies at the three cave bear species *U. s. ingressus* und *U. s. spelaeus* and *U. s. eremus* are presented on the Herman's Cave material (Fig. **25**). Dental and postcranial cave bear pathologies have been identified with first examples *e.g.* on material from the Czech Sloup Cave or Austrian Mixnitz Cave [36, 37]. In several publications about cave bears, single or few pathological finds were figured such as from the Perick Caves [21], the Dechen Cave [36, 37] and even the Hermann's Cave [6]. The problems of the former descriptions are the incorrect or do not attribute to the recent valid predator species. Furthermore, the historic and past descriptions did not take possibilities of top predator or cave bear fight trauma into account. Other problems occurred when the misidentifications of such inter and intraspecies fight trauma [18, 38 - 40] as proof for the "cave bear hunting by Neanderthals" [41].

Some isolated teeth show tooth root absorption due to inflammation or deformation of the enamel (Figs. **25A-F**). Tooth root inflammation is also visible on some jaws, in which alveoli are extended and surrounded by bone growth or dissolution (Figs. **25G-I**). A mandible of an older cub seems to have a trauma fracture, which was partly healed whereas all M1-3 teeth finally were lost

(Fig. **25I**). Such cranial jaw trauma might have been caused by interspecies predator or intraspecies fights. Interesting are the distal humerus bone dissolutions, which end in holes (those are no bite mark holes: Figs. **25J-L**). They develop either above the joint (Figs. **25J-K**), and in more rare cases laterally (looking like a felid humerus: Fig. **25L**). In most cases at the Hermann's Cave material the pathologies are periost reaction bone growth (= exostoses). Those seem to have two main causes:

1. Bite trauma reaching often the bone surfaces, for which the most impressive is a longer bite scratch on an ulna of a subadult cave bear (Fig. **25**). This one is quite wide in size and must have been caused by a large predator or other cave bear canine tooth. There are several bone growths on the ulna and radii (Figs. **25M-R**) or other pedal bones (Figs. **25X-Y**), which seem to be all caused by bite trauma. Cave bears battles were similar to modern bears with their fore leg paws, therefore injuries happened more often on the fore legs.

2. Arthritis mostly in elderly individuals. In the latter age class, syndesmophytes are common on the ventral discs/centra, which start to connect vertebrae (Figs. **25U-W**). The main zones for those are the cervical and lumbar areas, but in some cases also the anterior thoracic vertebrae can be affected (Fig. **25W**). Other bone growth are found around broken bones, which were in the stage of healing when the bears died (Fig. **25Z**).

The recent publications about steppe lion pathologies and Ice Age spotted hyena pathologies and similar cranial bite trauma damages supported the cave bear killer and battle in the cave theme [18, 22 - 25]. More and more pathologies can be corrected to be not of "rheumatic diseases or arthritis due to cold cave climate" [36, 37], instead more and more will become clearer as results of inter- and intraspecies fights. In addition this was most recently demonstrated for the

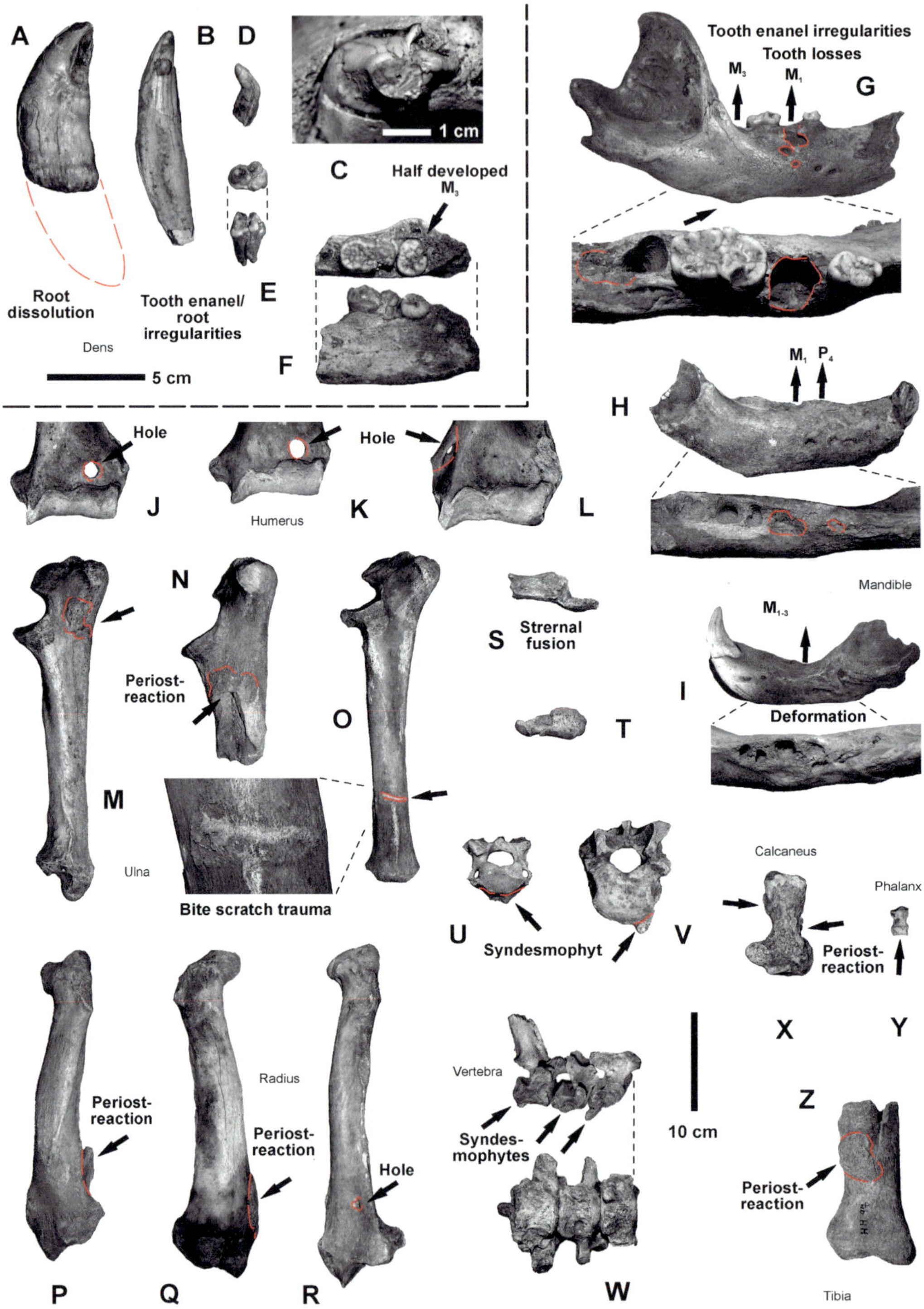
A
B
D
1 cm
C
Half developed
M3
Root
dissolution
Tooth enanel/
root
irregularities
E
Dens
F
5 cm
Tooth enanel irregularities
Tooth losses
M3
M1
G
M1
P4
H
Hole
J
K
Hole
L
Humerus
Mandible
M1-3
I
Deformation
N
Periost-
reaction
O
S
Sternal
fusion
T
M
Ulna
Bite scratch trauma
U
V
Syndesmophyt
Calcaneus
Phalanx
X
Periost-
reaction
Y
P
Q
R
Periost-
reaction
Radius
Periost-
reaction
Hole
Vertebra
Syndes-
mophytes
W
10 cm
Z
Periost-
reaction
Tibia

Figure 25. Cranial, and postcranial cave bear pathologies. A. Canine of *U. s. ingressus* with basal root dissolution as result of pulpa inflammation, Lower Schwemm Cave, Bear Hall (coll. RH). B-E. Too slim and on the enamel deformed canine of an *U. s. ingressus* cub, Bear Hall (coll. RH). E. P4 of *U. s. ingressus* with tooth cusp deformation, Bear Hall (coll. LMNB). F. Mandible of a cave bear cub with only half developed M3, Saal Hall (coll. LDA). G. Mandible with lost M1 und 3 of an adult cave bear (inflammation of the pulpa at M1), Saal Hall (coll. LDA). H. Mandible of an elderly cave bear with rounded canine root and nearly closed P4/M1 alveoli, Lower Fluvial Passage conus (coll. RH). I. Mandible of a cub with lost M1-3 and partly closed alveoli such as jaw deformation, Saal Hall (coll. LDA). J. Distal humerus (*U. spelaeus* subsp.) with hole pathology, Hermann's Cave (coll. RH). K. Distal humerus (*U. spelaeus* subsp.) with hole pathology, Hermann's Cave (coll. RH). L. Distal humerus (*U. spelaeus* subsp.) with lateral hole pathology, Hermann's Cave (coll. RH). M. Ulna (*U. s. ingressus*) with periost exostoses reaction, Hermann's Cave (coll. RH). N. Ulna of a subadult (*U. s. ingressus*) with periost exostoses reaction as result of large canine bite scratch trauma, Hermann's Cave (coll. RH). O. Ulna (*U. s. ingressus*) with periost exostoses reaction, Lower Schwemm Cave conus (coll. RH). P. Radius (*U. s. ingressus*) with periost exostoses reaction, Bear Hall (coll. RH). Q. Radius (*U. s. ingressus*) with periost exostoses reaction, Bear Hall (coll. RH). R. Radius (*U. s. ingressus*) with hole, Bear Hall (coll. RH). S. Sternal bone with exostoses, Saal Hall (coll. LDA). T. Second costa joint with exostoses, Bear Hall (coll. RH). U. Middle cervical vertebra with syndesmophyte, Bear Hall (coll. RH). V. Posterior thoracic vertebra (*U. s. ingressus*) with syndesmophyte, Bear Hall (coll. RH). W. Three anterior thoracic vertebra (*U. s. ingressus*) with syndesmophytes, Bear Hall (coll. RH). X. Calcaneus (*U. s. ingressus*) with periost exostoses reaction, Hermann's Cave (coll. LMNB). Y. Phlalanx II (*U. s. ingressus*) with joint deformation, Bear Hall (coll. RH). Z. Tibia with periost exostoses reaction, Bear Hall (coll. RH).

smallest cave bear species *U. s. eremus* from the Sophie's Cave [4]. There and in many cases the pathologies are bone growths on bone surfaces (periost reaction), that can result from different cases. Finally, also the cave bear hunt of the Late Palaeolithics, which is accepted and has been supported by several European cave sites during the Aurignacien-Gravettien at minimum [see Chapter Aurignacien cave bear hunters], must be taken into account for cave bear trauma injuries.

REFERENCES

[1] Grotrian H. Über das Vorkommen von Knochenresten der quartären Säugethier-Fauna in den Höhlen und Spalten des Devonkalkes von Rübeland am Harz. Z dt Geol Ges 1878; 30: 552-3.

[2] Kloos JH, Müller M. Die Hermannshöhle bei Rübeland. Weimar: K. Schwier-Verlag 1889; p. 76.

[3] Diedrich C. Evolution, Horste, Taphonomie und Prädatoren der Rübeländer Höhlenbären, Harz (Norddeutschland). Mitt Karst- Höhlenf 2013; 59(1): 4-29.

[4] Diedrich C. Sophie's Cave - a Late Pleistocene cave bear den. Bentham 2015; p. 192.

[5] Völker R. Die Grabungen auf Bärenknochen im Frühjahr 1984 in der Hermannshöhle Rübeland. Berichte und Mitt Zentr Fach Höhlen- Karstf 1985; 85(1): 5-7.

[6] Fischer K. Der Höhlenbär Ursus spelaeus Rosenmüller and Heinroth 1794 von Rübeland. Braunschweiger naturk Schr 1995; 4(4): 709-42.

[7] Diedrich C. Impact of the German Harz Mountain Weichselian ice-shield and valley glacier development onto Palaeolithics and megafauna disappearance. Quat Sci Rev 2013; 82: 167-98. [http://dx.doi.org/10.1016/j.quascirev.2013.08.013]

[8] Diedrich C. Ichnological and ethological studies in one of Europe's famous bear den in the Urşilor Cave (Carpathians, Romania). Ichnos 2011; 18(1): 9-26. [http://dx.doi.org/10.1080/10420940.2011.552578]

[9] Rabeder G, Nagel D, Pacher M. Der Höhlenbär Thorbecke Species 4. Stuttgart: Thorbecke 2000; p. 111.

[10] Rosendahl S, Döppes D. Trace fossils from bears in caves of Germany and Austria. Scient Ann School Geol Arist Univ Thessaloniki 2006; 98(spec): 241-9. [AUTH].

[11] Rabeder G. Die Evolution des Höhlenbärgebisses. Mitt Komm Quartärf Österr Akad Wissensch 1999; 11: 1-102.

[12] Stiller M, Baryshnikov G, Bocherens H, *et al.* Withering away--25,000 years of genetic decline preceded cave bear extinction. Mol Biol Evol 2010; 27(5): 975-8. [http://dx.doi.org/10.1093/molbev/msq083] [PMID: 20335279]

[13] Münzel SC, Stiller M, Hofreiter M, Mittnik A, Conard NJ, Bocherens H. Pleistocene bears in the Swabian Jura (Germany): Genetic replacement, ecological displacement, extinctions and survival. Quat Int 2011; 245: 225-37. [http://dx.doi.org/10.1016/j.quaint.2011.03.060]

[14] Pacher M, Stuart AJ. Extinction chronology and palaeobiology of the cave bear (*Ursus spelaeus*). Boreas 2008; 38(2): 189-206. [http://dx.doi.org/10.1111/j.1502-3885.2008.00071.x]

[15] Rabeder G, Hofreiter M, Nagel D, Paabo S, Withalm G. Die neue Taxonomie der Höhlenbären. Abh Karst- Höhlenk 2002; 34: 68-9.

[16] Rabeder G, Hofreiter M. Der neue Stammbaum der Höhlenbären. Die Höhle 2004; 55: 1-19.

[17] Hofreiter M. Genetic stability and replacement in late Pleistocene cave bear populations. Abh Karst- Höhlenk 2002; 34: 64-7.

[18] Diedrich C. Holotype skulls, stratigraphy, bone taphonomy and excavation history in the Zoolithen Cave and new theory about Esper's "great deluge". Eiszeit Gegenw. Quat Sci J 2013; 63(1): 78-98.

[19] Steiner U, Steiner W. Zur Genese der Rübeländer Höhlen. Geologie 1969; 18: 198-203.

[20] Kaulich B, Rosendahl W. The Neonate cave bear skeleton from the Petershöhle near Velden (Franconian Alb, Germany). Abh Karst- Höhlenk 2002; 34: 12-6.

[21] Dreyer R, Graw R. Die Bären im Sauerland: Knochenfunde aus der Dechenhöhle bei Iserlohn. Abh Karst- Höhlenk 2002; 34: 30-5.

[22] Diedrich C. Cave bear killers, scavengers between the Scandinavian and Alpine Ice shields – the last hyenas and cave bears in antagonism – and the reason why cave bears hibernated deeply in caves. Stalactite 2009; 58(2): 53-63.

[23] Diedrich C. Cave bear killers and scavengers from the last Ice Age of central Europe: Feeding specializations in response to the absence of mammoth steppe fauna from mountainous regions. Quat Int 2011; 255: 59-78. [http://dx.doi.org/10.1016/j.quaint.2011.06.048]

[24] Diedrich CG. 'Neanderthal bone flutes': simply products of Ice Age spotted hyena scavenging activities on cave bear cubs in European cave bear dens. R Soc Open Sci 2015; 2(4): 140022. [http://dx.doi.org/10.1098/rsos.140022] [PMID: 26064624]

[25] Diedrich C. Cracking and nibbling marks as indicators for the Upper Pleistocene spotted hyaena as a scavenger of cave bear (*Ursus spelaeus* Rosenmüller 1794) carcasses in the Perick Caves den of Northwest Germany. Abh Naturhis Ges Nürnberg 2005; 45: 73-90.

[26] Diedrich C. When European Ice Age spotted hyenas crushed their own to pieces. Roy Soc Open

[27] Diedrich C. Late Ice Age wolves as cave bear scavengers in the Sophie's Cave of Germany – extinctions of cave bears as result of climate/habitat change and large carnivore predation stress in Europe. ISRN Zool 2013; 2013: 1-25. [http://dx.doi.org/10.1155/2013/138319]

[28] Diedrich C. Late Pleistocene Eemian hyena and steppe lion feeding strategies on their largest prey - *Palaeoloxodon antiquus* Falconer and Cautley 1845 at the straight-tusked elephant graveyard and Neanderthal site Neumark-Nord Lake 1, Central Germany. Archaeol Anthropol Sci 2013; 6(3): 271-91.
[http://dx.doi.org/10.1007/s12520-013-0150-7]

[29] Diedrich C. Whoolly mammoth carcass scavengers – Ice Age spotted hyenas and Steppe lions along the Central European Rhine Graben "Ice Age Safari" big game migratory channel. Quat Int

[30] Diedrich C. The Late Pleistocene spotted hyena *Crocuta crocuta spelaea* (Goldfuss 1823) population from the Zoolithen Cave at Gailenreuth (Bavaria, South Germany) – a hyena cub rising den of specialized cave bear scavengers in Boreal Forest environments of Central Europe. Hist Biol 2011; 23(4): 335-67.
[http://dx.doi.org/10.1080/08912963.2011.552719]

[31] Diedrich C. The Late Pleistocene *Crocuta crocuta spelaea* (Goldfuss, 1823) population from the Emscher River terrace hyena open air den Bottrop and other sites in NW-Germany – woolly rhinoceros scavengers and their bone accumulations along rivers in lowland mammoth steppe environments. Quat Int 2012; 276/277: 93-119.
[http://dx.doi.org/10.1016/j.quaint.2011.07.046]

[32] Stiner MC. Mortality analysis of Pleistocene bears and its paleoanthropological relevance. J Hum Evol 1998; 34(3): 303-26.
[http://dx.doi.org/10.1006/jhev.1997.0198] [PMID: 9547458]

[33] Grandal-D' Anglade A, Vidal Romani JR. A population study of the cave bear (*Ursus spelaeus* Ros.-Hein.) from Cova Eiros (Triacastela, Galicia, Spain). Geobios 1997; 30(5): 723-31.
[http://dx.doi.org/10.1016/S0016-6995(97)80160-X]

[34] Debeljak D. Fossil population structure and mortality of the cave bear from the Mikrica Cave (North Slovenia). Acta Carsol 2007; 36(3): 475-84.
[http://dx.doi.org/10.3986/ac.v36i3.181]

[35] Argenti P, Mazza PP. Mortality analyses of the Late Pleistocene bears from Grotta Lattaia, central Italy. J Archaeol Sci 2006; 33: 1552-8.
[http://dx.doi.org/10.1016/j.jas.2006.02.007]

[36] Breuer R. Zur Anatomie, Pathologie und Histologie der Zähne und Kiefer von *Ursus spelaeus*. Spelaeogr Monogr 1931; 7(8): 581-610.

[37] Breuer R. Pathologisch-anatomische Befunde am Skelett des Höhlenbären. Spelaeogr Monogr 1931; 7-8: 611-23.

[38] Diedrich C. The largest European lion *Panthera leo spelaea* (Goldfuss) population from the Zoolithen Cave, Germany – specialized cave bear predators of Europe. Hist Biol 2011; 23(2-3): 271-311.
[http://dx.doi.org/10.1080/08912963.2010.546529]

[39] Diedrich C. Late Pleistocene *Panthera leo spelaea* (Goldfuss, 1810) skeletons from the Czech Republic (central Europe); their pathological cranial features and injuries resulting from intraspecific fights, conflicts with hyenas, and attacks on cave bears. Bull Geosci 2011; 86(4): 817-40.
[http://dx.doi.org/10.3140/bull.geosci.1263]

[40] Rothschild BM, Diedrich C. Comparison of pathologies in the extinct Pleistocene Eurasian steppe lion *Panthera leo spelaea* (Goldfuss, 1810) to those in the modern lion, *Panthera leo* – Results of fights with hyenas, bears and lions and other ecological stress. Int J Paleopathol 2012; 2(4): 187-98.
[http://dx.doi.org/10.1016/j.ijpp.2012.09.004]

[41] Groiss JT. Über pathologische Bildungen an Skelettresten jungquartärer Säugetiere aus der Zoolithenhöhle bei Burggeilenreuth. Geol Bl NO-Bayern 1978; 28(1): 1-21.

CHAPTER 4

Steppe Lions - Cave Bear Killers

Abstract: In the large northern German Hermann's Cave, one skull and 127 bones of four lion skeletons have been excavated since 1968 which are presented in an osteologi monograph. Three of them are incomplete and were found scattered in a smaller area (3 m2) in autochthonous Late Pleistocene MIS 3-5d layers at a distance of 90 far from the former entrance. It is expected that further bones completing those in future digs. Those three originally articulated *Panthera leo spelaea* (Goldfuss, 1810) (popular name = Ice Age steppe lion) few scattered skeleton remains of early adult to grown up individuals were found nearby the end of the 90 meters deep into the middle ponor level reaching horizontal cave bear den cave. Those lions represent an elderly-pathological (dental pathology – all broken and polished canine teeth, broken rib with callus), adult male and two adult female lions. A new and updated bone metric statistics (crania and longbones) including material from the Upper Rhine Graben (southern Germany) demonstrates a size increase of the lions according to the Bergmann's Rule. The Eemian lions are small, increase in size with the temperature drop and have finally largest forms (reching even nearly similar in size to the older *P. l. fossilis* Reichenau, 1906) in the cold MIS 3 period, before the Last Glacial maximum, when they got extinct. Three variants are distinguished with the small Eemian Interglacial (MIS 5e) *P. l. s. brachygnathus* Dietrich, 1968, (holotype mandible from Dechen Cave and paratype skeleton of neumark Nord Lake 1, Germany) the most common Weichselian/Wuermian (MIS 4-5d) *P. l. s. intermedia* nov. var. (holotype skull and all herein described material from the Hermann's Cave) and the largest form *P. l. s.* var. *maximus* (holotype skeleton from Huttenheim, Germany). Originally articulated skeletons of one to four adult lion individuals where found further from entrances in easy accessible larger cave bear dens are found in at 14 European cave bear den sites most convincing for the Urşilor Cave (Ro), the Sloup Cave (CZ), the Bilstein Cave (D), and herein added, in the Hermann's Cave (D). Those sites prove cave bear predation by the last lions of Europe in boreal forest mountain regions, most probably accured mainly in winter times during cave bear hibernation and in spring, for cub killing even most far into the cave system. At several lion remains from cave sites large canine (unclear origin from cave bears, lions or hyenas) bite damages on skulls or fewer longbone joints such as postcranial trauma pathologies indicate "battles" with "large prey/antagonists" most probably in the caves. Natural death by trapping of those nocturnal climber felids can be excluded in nearly all cases, because most of them are horizontal (ponor-level) caves. In rare cases, trapping of lions in complex and large and vertical connected cave systems cannibalism might have happened under stress situation found with a possible example of one scavenged subadult lion in the Romanian Urşilor Cave. Lions must have died mainly during interspecies battles with smaller *Ursus spelaeus eremus* (= small cave bear) or large body sized *Ursus spelaeus ingressus* (= large cave bear). Lions are represented in the cave bear den caves only by 1-3% of the bone NISP. If only one

Cajus G. Diedrich

weak lion/lioness was killed per 25.000 years in a larger cave bear den, this would explain the low amount of lion skeletons (bones) due to rare lost battles, whereas signs of deathly injuries on bones of lions are absent, if only soft tissues were damaged causing mortality. Those lions were not consumed by the herbivorous cave bears, explaining why their articulated skeletons between cave bear nests and "carcass bonebeds".

Keywords: Late Pleistocene, Cave bear dens, Lion taphonomy, Articulated skeletons, Cave bear predation, Europe.

FOSSIL RECORD IN EUROPE AND GERMANY

The "cave lion" ("*Felis spelaea*") [1, 2], which is still in incorrect popular name use, now renamed as Ice Age steppe lion, was created based on a mixed skull and two different mandibles from the German Zoolithen Cave [3]. The holotype skull of *Panthera leo spelaea* (Goldfuss, 1823) was rediscovered after its historical disappearance over a time span of about 200 years [3 - 5].

The global coexistence of the Late Pleistocene European steppe lion *Panthera leo spelaea* with different Late Pleistocene lion subspecies was solved by DNA studies for the Late Pleistocene glacial record (MIS3-5d) [6, 7]. The boundary between European steppe lions and the East-Asian/Siberian/Beringian lions *Panthera leo vereshchagini* remains unclear in their exact Eurasian palaeobiogeography (Fig. **1A**) due to scarce finds in Siberia [7]. At the same time, in northern America *Panthera leo atrox* was present, and in Africa, a subspecies of the modern African lion *Panthera leo leo*, whereas Ice Age tigers appeared only in Asia [7].

The Late Pleistocene *P. l. spelaea* appeared in Central Europe (and Germany) after the climate change with the glacial "Mammoth-fauna" about 113.00 years ago in the MIS5d, early Late Pleistocene [6 - 8]. High elevated remains were found in Europe *e.g.* about 1.600 m a.s.l. in alpine regions, such as at the Wildkirchli Cave cave bear den near Appenzell (Switzerland, Fig. **1B**) [9]. Those lions are absent in higher elevated alpine conditions where Ice Age leopards instead filled the ecological niche [10].

Mainly isolated teeth and postcranial bones have been described and figured from Europe mostly from caves and in lower amounts at open air sites [1 - 41]. *P. l. spelaea* is recently known from Europe by 35 mostly complete skulls which were compiled and reviewed [8]. A fairly complete skeleton of an subadult lioness with braincase damage was described from the Late Pleistocene glacial Srbsko Chlum-Komín Cave hyena den (Czech Republic) [34] (Fig. **1D**). Remains of three Late Pleistocene glacial skeletons from the Urşilor Cave cave bear den (Ro) are the

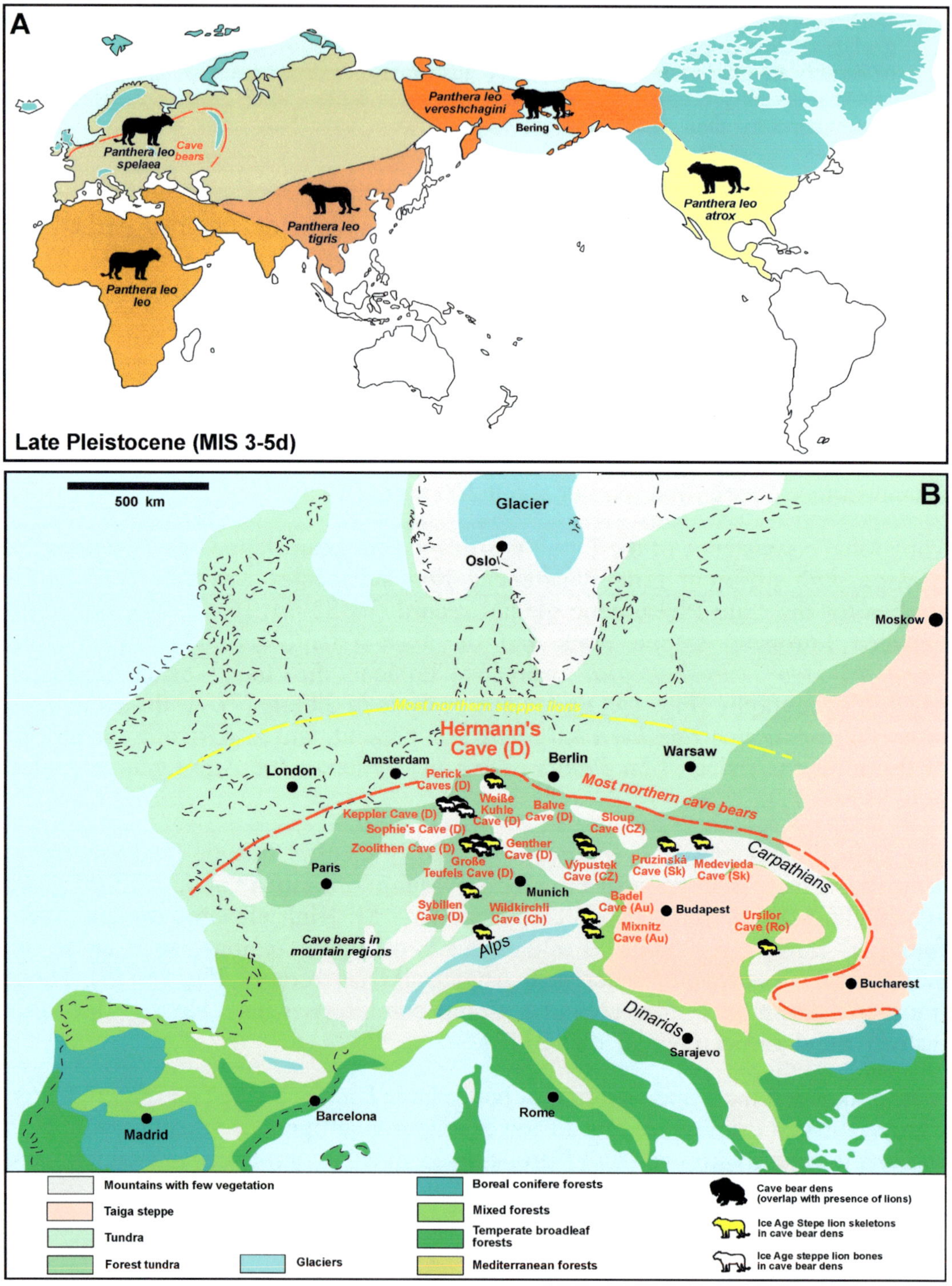
A
Panthera leo spelaea
Cave bears
Panthera leo vereshchagini
Bering
Panthera leo tigris
Panthera leo leo
Panthera leo atrox
Late Pleistocene (MIS 3-5d)
B
500 km
Glacier
Oslo
Moskow
Most northern steppe lions
Hermann's Cave (D)
London
Amsterdam
Berlin
Warsaw
Most northern cave bears
Perick Caves (D)
Weiße Kuhle Cave (D)
Balve Cave (D)
Keppler Cave (D)
Sophie's Cave (D)
Sloup Cave (CZ)
Zoolithen Cave (D)
Genther Cave (D)
Große Teufels Cave (D)
Výpustek Cave (CZ)
Pruzinská Cave (Sk)
Medevieda Cave (Sk)
Carpathians
Paris
Munich
Sybillen Cave (D)
Wildkirchli Cave (Ch)
Badel Cave (Au)
Budapest
Mixnitz Cave (Au)
Ursilor Cave (Ro)
Cave bears in mountain regions
Alps
Bucharest
Dinarids
Sarajevo
Madrid
Barcelona
Rome
Mountains with few vegetation
Taiga steppe
Tundra
Forest tundra
Glaciers
Boreal conifere forests
Mixed forests
Temperate broadleaf forests
Mediterranean forests
Cave bear dens (overlap with presence of lions)
Ice Age Stepe lion skeletons in cave bear dens
Ice Age steppe lion bones in cave bear dens

Figure 1. A. Global Late Pleistocene lion subspecies distribution (after DNA data from [6, 7], palaeogeographic map after [8]). B. Cave bear den sites with Ice Age steppe lion *Panthera leo spelaea* (Goldfuss, 1823) skeleton remains (or several bones of skeleton parts) in Europe. Those often overlap with hyena den sites (as imported prey) or are found in cave bear dens (as demonstrated herein because of lost battles during cave bear hunt).

most spectacular lion finds ever found more than one km from the former entrance deep in a cave [40, 41]. Other skeleton remains came from Czech Republic (Praha-Podbaba) or German river terrace open air sites [15, 26, 27, 32]. Additional, lion skeletons have also been published in unclear taphonomic context from larger caves in Spain (Arrikrutz) [12] and Austria (Salzburg) [13].

From Germany, the amount of steppe lion *P. l. spelaea* bones and skeletons and their sites have increased enormously since 2005 due to the authors research about the "top predators of the Ice Age", mainly hyenas and lions of the last glacial [8]. A number of bones of the former believed "rare lion" were simply overlooked in many cave bear den sites and in several museums collections. Especially the postcranial bones (*e.g.* radius, vertebrae and ribs), which look in some cases indeed very similar to those of cave bears, often which were found in museum's collections between cave bear bones (*e.g* [19, 34].). There are now five by the author main studied regions in Germany (Fig. **3**): 1. Open air sites around Berlin [32], 2. The Sauerland Karst cave-rich region [28, 29, 33], 3. The Upper Rhine Graben and southern German caves [36], 4. The Upper Franconia caves [3 - 5, 35], and the herein described, 5. Harz Mountain range (Baumann's/Hermann's and Unicorn Cave) and surrounding open air such as the Thuringian open air and cave sites (*e.g.* Westeregeln, Thiede). About 1.500 steppe lion bones are now described from German localities including the material of the Hermann's Cave (herein 128 bones).

Mainly isolated teeth, a few skulls, and postcranial bones, especially the easy to identify bones such as the humerus with its supracondylar foramen, straight and in middle shaft cross-section nearly round femur, or typical metapods with round distal joints, were described and figured in many publications for German sites [2 - 5, 16, 28, 29, 32, 35]. The only known and very fragmentary young cub skeleton was found within a hyena den (Wilhelms Cave, D) [33]. It seems to have been imported as "hyena prey" or was killed by hyenas/or cave bears at their den site during prey stealing or cub killing purposes of lions [4]. Late Pleistocene glacial adult lion skeletons (none of those are complete) have been found at different German open air river terrace sites: Siegsdorf, Huttenheim, Edingen, and Brühl (Fig. **2**) [15, 36]. Another skeleton with a more recent heavily damaged pelvic area of another sickly lioness was found within the German Neumark-Nord Lake 1 open air "elephant graveyard" and Neanderthal camp site dating to the Eemian Interglacial [30].

North Sea
Dk
100 km
P. l. spelaea bones
P. l. spelaea skeletons
Open air sites (hyena influenced)
Cave sites (hyena dens)
C. c. spelaea bones
GERMANY
Nl
Pl
Brandenburg
Lower Saxony
Saxony-Anhalt
Berlin
Study areas
Westphalia
Perick Caves
Thiede
Westeregeln
Hermann's Cave
Wilhelms Cave
Keppler Cave
Altmorschen
Lindenthal Cave
Hesse
CZ
Zoolithen Cave
Sophie's Cave
Große Teufels Cave
Huttenheim
Brühl
F
Bavaria
Siegsdorf
A
Baden-Wuerttemberg

Figure 2. Late Pleistocene steppe lion *Panthera leo spelaea* (Goldfuss, 1823) skeletons, skulls and isolated bones in Germany. In many cases those overlap with hyena dens and are imported carcass remains. (Composed from the author's publications [8, 28 - 30, 32, 33, 35, 36, 44]).

HABITUS RECONSTRUCTION

The Ice Age steppe lion (former "cave lion") was several times engraved or painted in the Late Palaeolithic Aurignacien to Magdalenian European cave/mobile art between 40.000-16.000 years ago. The two most important figurations have anatomical details such as the mane and only for lions the typical black tassel [42, 43] (Fig. **3**).

The clearest painting of the black tassel in the French Chauvet Cave (Fig. **3B**) and the mane in the "en-face" lion engraving in Trois Frères Cave (Fig. **3C**) [11, 12] confirm, that those were lions and not tigers, of which of the later have no tassel. Furthermore they show that males probably similar to modern African lions, a mane [36]. The combination of the Palaeolithic art, skeleton anatomy and osteometric data [8, 36] being added herein with new skeleton material (see Table **1**), allow an updated 3D presentation of this European Late Pleistocene steppe lion. Those were only about 1/4 larger as modern African lions [15, 36]. The largest lion in the European Ice Age was *Panthera leo fossilis* which lived during the Middle Pleistocene [25]. The fur colour or fur colour change from cubs to adults remains speculative, but is based on the life of open environment grassland adapted felids such as extant lions (see Fig. **3A**).

HERMANN'S CAVE STEPPE LIONS

Skeleton Distribution in the Cave and Carcass Taphonomy

The position of the skeleton remains (Fig. **4**) can be reconstructed by the excavation reports, whereas only the bones from the Voelker excavation 1984 are mentioned to have been mapped in a general bone mapping without distinguishing those from cave bear bones (see Chapter Introduction). Sadly, this report is neither available nor published to reconstruct their exact position within the excavation area. By the reports of Steiner and Voelker [44, 45], both excavations were made adjacent to each others covering about 4-6 square meters in the eastern part close to the cave wall of the Saal Hall (Fig. **4**). In both excavation squares the main lion material was found in the upper autochthonous cave bear clays in-between the abundant cave bear remains. The large amount of bones of three skeletons in such a small area obviously results of originally articulated skeletons, which were scattered. Furthermore, it is obvious after this study herein, that those skeletons can be completed by new excavations in the Saal Hall.

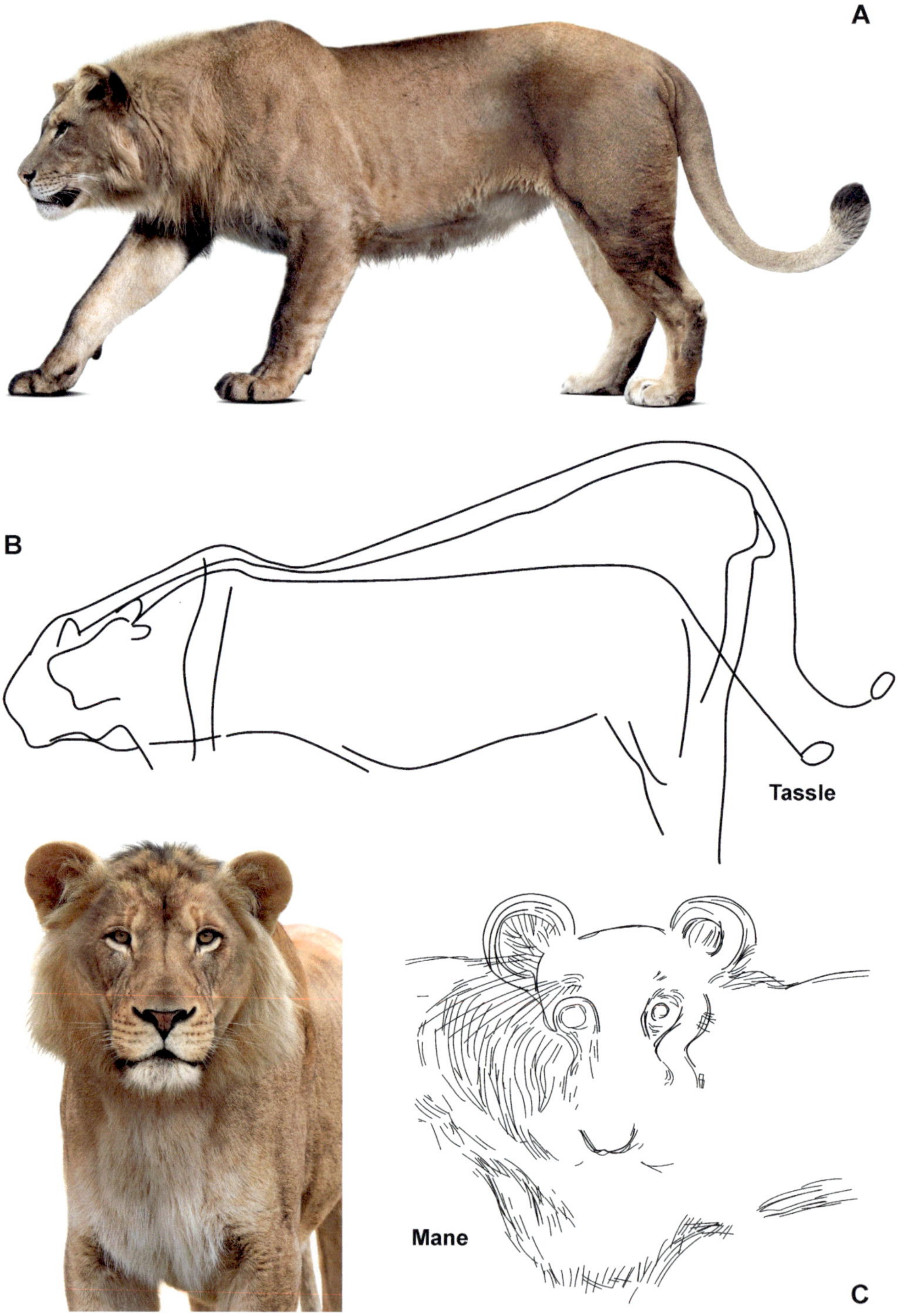
A
B
Tassle
Mane
C

Figure 3. Late Pleistocene *Panthera leo spelaea* (Goldfuss, 1823) = Late Pleistocene Ice Age steppe lion. A. Model of an early adult male (lion illustrations R. Uchytel). B. Outline of the Aurignacien/Gravettien paintings in Chauvet Cave (France) of lionesses with presence of the tail tassel typical for lions (redrawn from [43]). C. En-face engraving of a male lion with a mane in the Trois-Frère Cave (France) (redrawn from [42]).

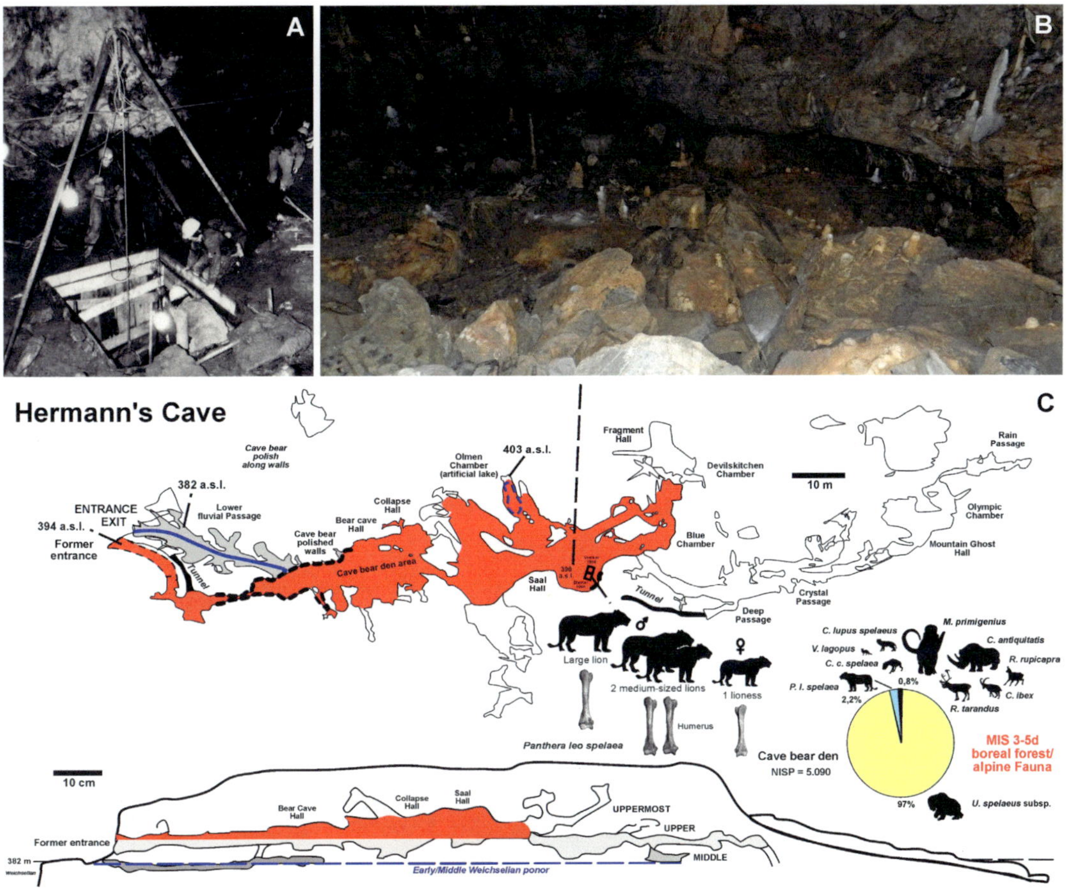

Figure 4. Hermann's Cave cave bear den with four lion skeleton remains which three Elderly-adults were found all about 90 meters away from the former entrance in the Saal Hall (at the end of the den), subadult remains might be in the context of a short-term hyena den site at the former entrance (= imported with other prey remains; modified from (modified from [35, 44]). A. Excavation claim of Voelker 1984, Right part of the Sall Hall with larger ceiling blocks and candle stalagmites. C. Cave map and percentages of the MIS3-5d fauna dominated by cave bears which includes only 2.2% NISP lion material.

All damages on bones are of modern excavation or trampling origin. There are no signs of any large predator scavenging. Bite marks or bone damage stages as result of intensive carnivore feeding are also absent. Possibly, there are few signs of smaller carnivore damage distally on some dorsal spines, whereas their breakages might also have other origins. A scavenging of all four skeletons by hyenas, wolves or even cave bears can be excluded if compared to hyena or lion damaged skeletons from *e.g*. the Urşilor Cave (cannibalistic lion damage [8, 41, 47]) or Perick Caves (hyena damage [29]).

Dating of the Skeletons/Bones

The geology of the region, sedimentary refill of the cave systems at Rübeland and their bone taphonomy was analyzed in a first step for both Rübeland caves, the Baumann's and the opposite Hermann's Cave [46]. A more detailed discussion to this can be found in the Chapter Geology. The oldest sediments in the Hermann's Cave are found in the middle ponor level (Fig. **4**) [46]. Middle Pleistocene Bode River terrace gravels and sands were washed into the Hermann's Cave during the late Saalian Glacial (Figs. **4**, **5**) [46]. Over this, autochthonous Late Pleistocene cave sediments with red cave clay and corroded and fallen limestone blocks built the cave bear and lion bearing bone-rich autochthonous layers which contain even articulated cave bear skeletons (see Chapter cave bears) [46]. The lion remains from the Hermann's Cave (Saal Hall) are from the upper 50 cm of the section. The youngest speleothem generation and candle stalagmites covered the *Ursus spelaeus ingressus*, *Panthera leo spelaea* bones and archaeological Aurignacian finds in the Hermann's Cave [46].

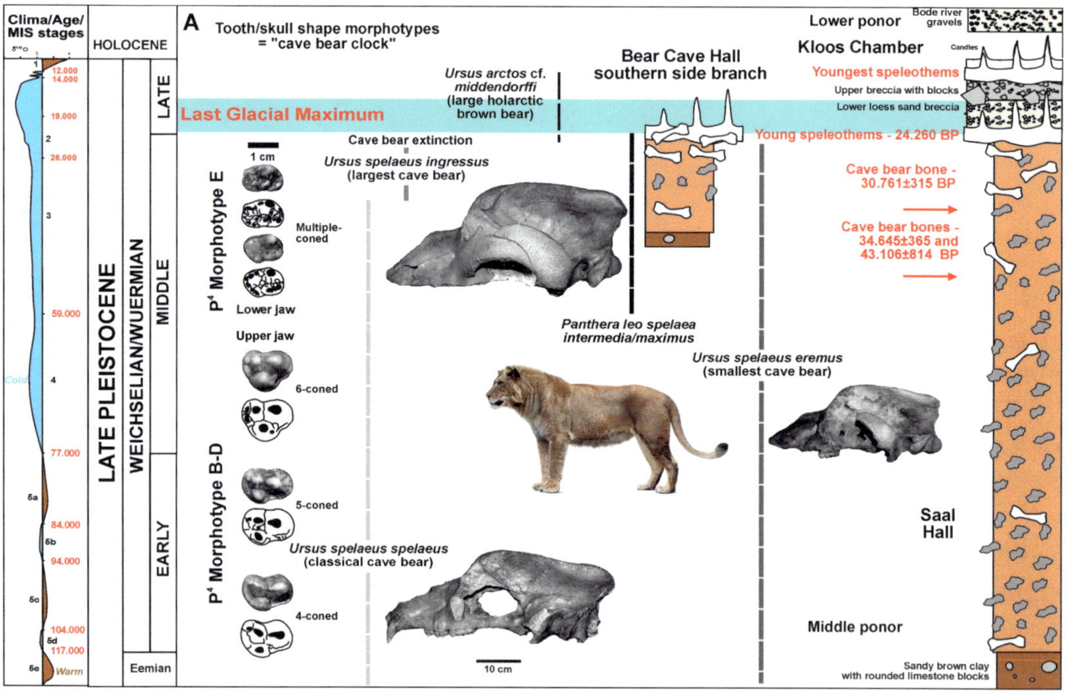

Figure 5. Sedimentary stratigraphy and dating of the lion skeletons which were found in the upper part of the section of the Saal Hall in the Hermann's Cave (modified from [35, 46]).

Its covering speleothem layer date all those remains including the lions, to be older than 23.130 ± 711 and 24.260 ± 568 BP [46, 48] (Fig. **5**). The lion skeletons can be dated, also using the "cave bear clock" (see Chapter cave bears), between approximately 130.000-24.000 BP into the Late Pleistocene glacial (= MIS 3-5d). During this, and the following Lass Glacial Maximum (=LGM)-time, the lower ponor level developed due to melting valley glacier waters [46]). Finally, in the Post-LGM, the former entrance was closed by frost breccias, which fallen larger blocks also covered larger parts of the "bonebed" layer all over the Hermann's Cave (Fig. **5**) [46].

Skeleton Material

All lion "bones" from the Saal Hall and bones mixed into the former "bone dump" were overlooked during the Steiner 1968 excavations [44]. A quite large amount of postcranial bones was excavated and is herein identified for the first time being separated now from the mixed cave bear material (see Table **1**). Similarly overlooked were the excavated lion remains from the Voelker excavation in 1984 [cf. 45]. Initial identifications and descriptions of parts were presented by Fischer, but the skull was not identified nor fixed [16]. Also ribs or other postcranial bones such as several vertebrae were overlooked between the cave bear bone material. The lion patellae were confused in the publication as "pointed form" with those of cave bears [16]. The authors own cave survey in 2011 resulted at the Voelker 1984 excavation area in some smaller bone dumps some mainly damaged lion remains, also in the bone dump of the cave bear hall [46]. One skull, one mandible and 126 bones from the two excavations and one prospection can be figured with all material from four (mainly three) individuals, most of which are from the Saal Hall of the Hermann's Cave (Table **1**, Figs. **6-12**). In this stage of the limited material, not all bones can be clearly attributed to each skeleton, especially the thoracic ones (ribs, sterna). Also vertebrae can be identified at the moment only to males (larger) or females (smaller). If future excavations will deliver the missing material, in addition the finds presented herein will be more clear in their separation to skeletons.

A minimum individual number (=MNI) of four grown up to elderly lions (= skeletons 1-4, or abbreviation Sk 1-4) can be calculated based on the humeri from the Sall Hall (Figs. **7.2-7.5**). Those allow bone proportions to be compared to other European material [cf. 4, 13-15, 30-34, 36] to distinguish between "three size classes", which seem to represent after the sex dimorphism of Late Pleistocene lions [36] compared to modern lions [15] males and females. In the

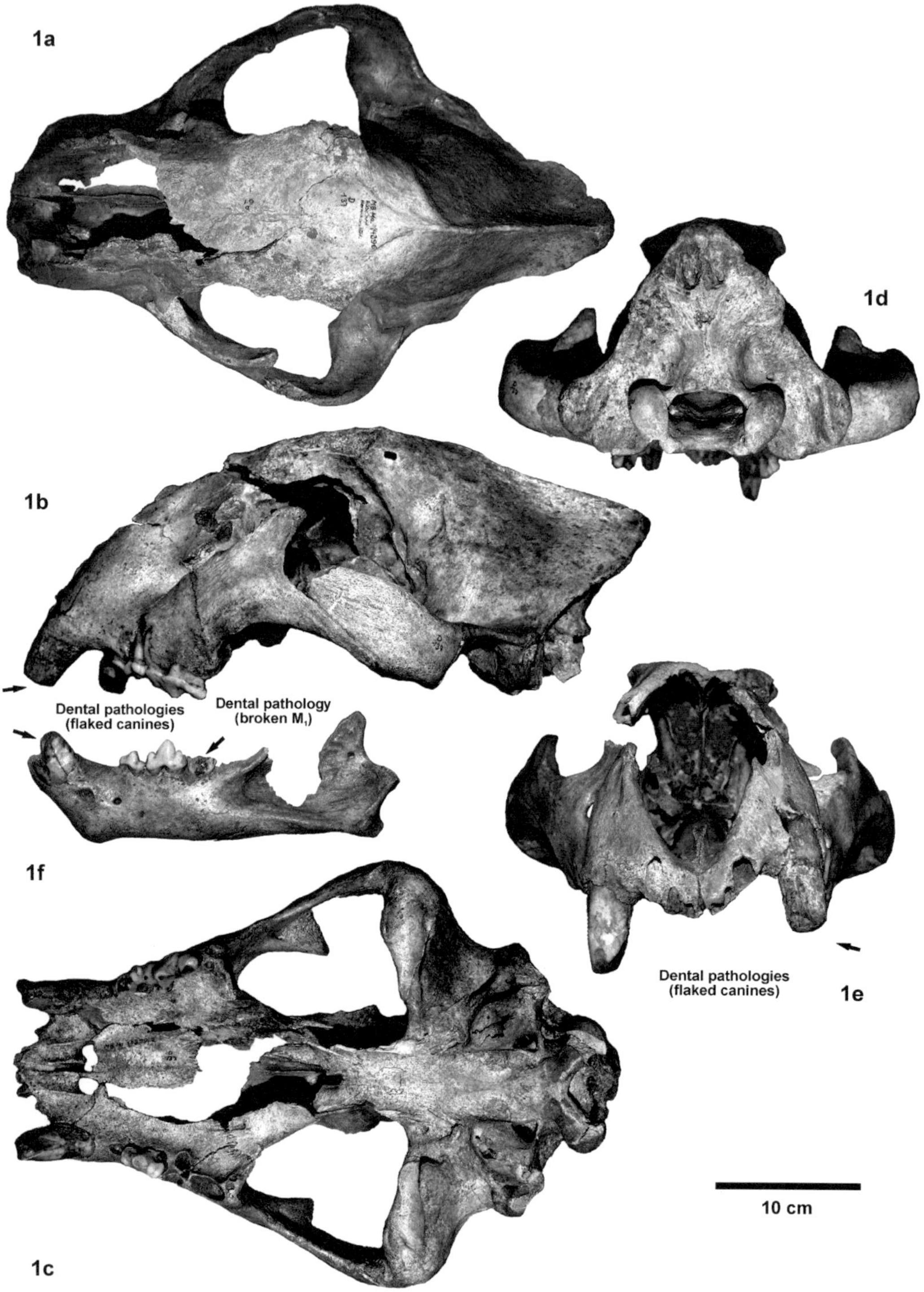
1a
1d
1b
Dental pathologies
(flaked canines)
Dental pathology
(broken M_1)
1f
Dental pathologies
(flaked canines)
1e
1c
10 cm

Figure 6. Cranium of the *Panthera leo spelaea* (Goldfuss, 1810) var. *intermedia* nov. var. skeleton 2, elderly pathological female individual of the Late Pleistocene (MIS 5a-d) Hermann's Cave, Rübeland (Saxony-Anhalt). a. dorsal, b. ventral, c. lateral left, d. mandible (skull: MB no. Ma.14299, mandible: MB no. Ma.14298 (metrics and see Table **1**, Fig. **12**).

Saal Hall there are remains of one large male and two smaller males, and one much smaller female. Based on the strong sexual dimorphism in larger males and smaller females [15, 36] the determinable material has a ratio of 71 males/15 female bones (Table **1**). This percentage correlates well to the humerus ratio of 3 males/1 female.

The studied lion bone material from the Hermann's Cave is stored in the following collections: Rübeland-Höhlen (= **RH**), Landesmuseum für Naturkunde Braunschweig (= **LMNB**), Museum für Naturkunde Berlin der Humboldt Universität (= **MB**), Landesmuseum Archäologie, Saxony-Anhalt (= **LDA**).

Skeleton 1 (= Sk1), largest lion - Only six bones can be composed more securely to this largest-sized adult-elderly male skeleton. The largest left humerus of Fig. **7.2** (cf. Fig. **12**) and one left femur seems to belong to this individual (Fig. **10.1**). To the massive femur, the most massive and in the shaft most wide rigth tibia (Fig. **10.5**) fits best to the same individual, whereas it falls in length also to the medium-sized lions (cf. Fig. **11**).

The large patella (= "pointed type of cave bear patella" in [16]), and one large calcaneus (Fig. **10.13**), such as the large astragalus (Fig. **10.16**) fit well into the femur/tibia joint sizes. The metapods are not clear in their identification to Sk 1-3. Possibly, some larger vertebrae might come from Sk 1, such as some ribs or sterna bones (Fig. **10**).

Skeleton 2/3 (= Sk2/), Medium-sized Male

The similar size to skeleton 3 makes it impossible to distinguish at the moment both, except the two humeri securely. One right humerus (Fig. **7.3**) and another nearly similar sized right humerus (Fig. **7.2**) represent Sk 2/3 at minimum of those two similar aged adult males. The ulna (Fig. **7.6**), and both radii (Figs. **7.7, 7.8**) are after comparison (Fig. **12**) medium-sized and might belong also to Sk 2/3. Some of the medium-sized matacarpi (Figs. **7.13, 7.16 -7.23**) are insecure to belong to this, or the Sk 2/3. Similar is the situation with the phalangae (Figs. **7.24 -7.30**). The axial skeleton of Sk1-3 is represented by several vertebrae (cf. Figs. **8, 9**), whereas all larger ones are impossible yet to identify securely to the Sk 1-3 male individuals. *E.g.*, the larger-sized cervical to lumbar vertebrae (Figs. **9.1, 9.2, 9.4, 9.5, 9.7-9.10**) might belong either to this Sk 2/3. Furthermore, the femorae (Figs. **11.2, 11.3**) or tibiae (Figs. **11.6-7**) such as distal fibulae (Figs. **11.9, 11.11**)

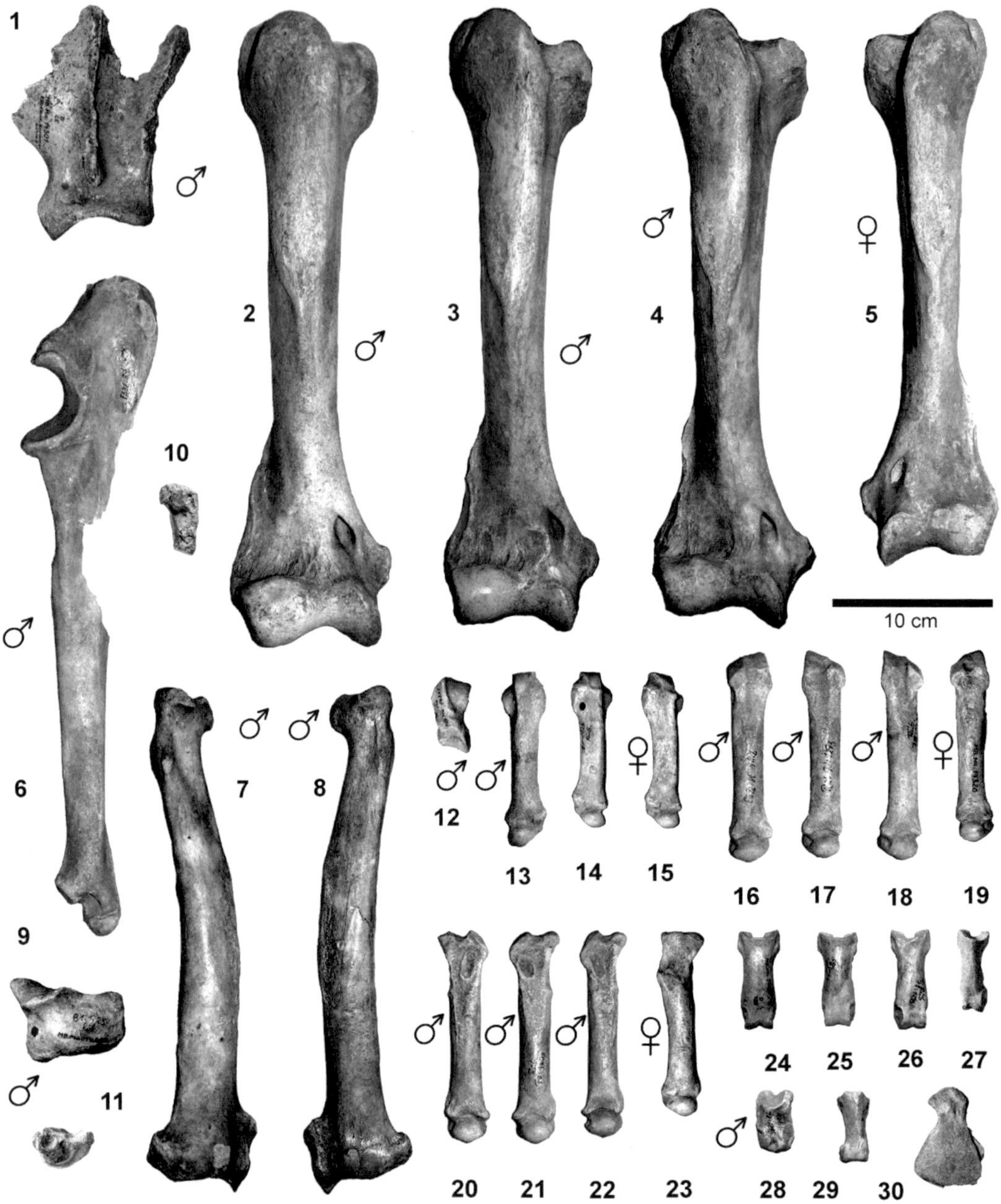

Figure 7. Postcranial fore limb bones of *Panthera leo spelaea* (Goldfuss, 1810) var. *intermedia* nov. var. from the Hermann's Cave, Rübeland, Saxony-Anhalt (Germany). 1. Left scapula (MB no. Ma.14301), lateral. 2. Right humerus (LDA no. 53:89), cranial. 3. Right humerus (MB no. Ma.14336), cranial. 4. Right humerus (MB no. Ma.14337), cranial. 5. Left humerus (LDA no. 78:1293a), cranial. 6. Left ulna (LDA no. 78:1293b), lateral. 7. Right radius (RCC no. HH-8), lateral. 8. Left radius (MB no. Ma.14332), lateral. 9. Left scapholunatum (MB no. Ma.14322), dorsal. 10. Right pisiform (MB no. Ma.14329), dorsal. 11. Right carpal

III (MB no. Ma.14326), lateral. 12. Left McII (MB no. Ma.14321), lateral. 13. Left Mc V (LDA no. 78:1293i), dorsal. 14. Right Mc V (MB no. Ma.14318), dorsal. 15. Right Mc V (MB no. Ma.14319), dorsal. 16. Left Mc IV (LDA no. 78:1293f), dorsal. 17. Right Mc IV (LDA no. 78:1293h), dorsal. 18. Right Mc IV (LDA no. 78:1293g), dorsal. 19. Left Mc IV (MB no. Ma.14320), dorsal. 20. Right Mc III (LDA no. 78:1293e), dorsal. 21. Left Mc III (LDA no. 78:1293d), dorsal. 22. Left Mc III (LDA no. 78:1293c), dorsal. 23. Left McII (MB no. Ma.14317), lateral. 24. Phalanx I of the right ?digit IV (MB no. Ma.9910), dorsal. 25. Phalanx I, right digit IV (LDA no. 78:1293zz), lateral. 26. Phalanx I, right digit II (LDA no. 78:1293j), lateral. 27. Phalanx I, right digit V (MB no. Ma. 9953), dorsal. 28. Phalanx I (LDA no. 78:1293k), lateral. 29. Phalanx II (LDA no. 78:1293l), lateral. 30. Phalanx III (LDA no. 78:1293m), lateral (metrics see Table **1**, Fig. **12**).

are again unknown to have originated from those skeletons. One intermediate sized astragalus (Fig. **14**) is another bone that fits to in proportions. As for the fore paw, the pedal bones are unclear to separate to the two medium-sized males, *e.g.* the metatarsi (Figs. **11.25, 11.27**) or finally all phalangae (Figs. **11.30, 11.34**).

Skeleton 4 (= Sk 4), Elderly Pathogenic Small Female

The only present cranium and left fitting mandible (Fig. **6**) belongs to an elderly individual. Their fits can be proven by the same tooth wear such as flake wear damages and mandible size. The strong fusion of the brain case and anterior sutures (maxillary, nasal) and especially the strongly used and use-flaked and further polished teeth including the canines are quite unique in the recent European lion record indicating a very old animal, maybe the oldest in individual age which has been found yet from the European Late Pleistocene [cf. 8]. The skull compared to other European sites is in smaller male proportions (Fig. **12**) [8]. At minimum 13 postcranial bones are quite obvious from the smallest proportioned lioness. The smallest humerus (Fig. **7.5**) is far out of the range compared to all three other ones (Fig. **12**). Other small-proportioned fore leg bones are the left Mc IV (Fig. **7.19**) and the left McII (Fig. **7.23**). The axial skeleton with its small proportioned vertebrae (especially much smaller centra) consists of six thoracic to lumbar (Figs. **8.8, 8.10, 8.11, 8.14, 8.16, 9.3**). Hind limb bones which can be identified to belong to females consist of a small proportioned proximally incomplete tibia (Fig. **11.8**), a distal fragment of the fibula (Fig. **11.10**), such as the smallest calcaneus (Fig. **11.12**) and smallest astragalus (Fig. **11.15**).

Lion Subspecies and Sizes: Sex Differences and Climatic Variants

The Hermann's Cave lion bone material, especially dentition of the skull and lower jaw was compared to the Mid-/Late Pleistocene lion record of different original descriptions and figures in publications about the German Cromerian (MIS 21-10) Mosbach lion *P. l. fossilis* [25, 37 - 39], which material will be revised in a separate publication, and the latest Middle Pleistocene/Late Plei-

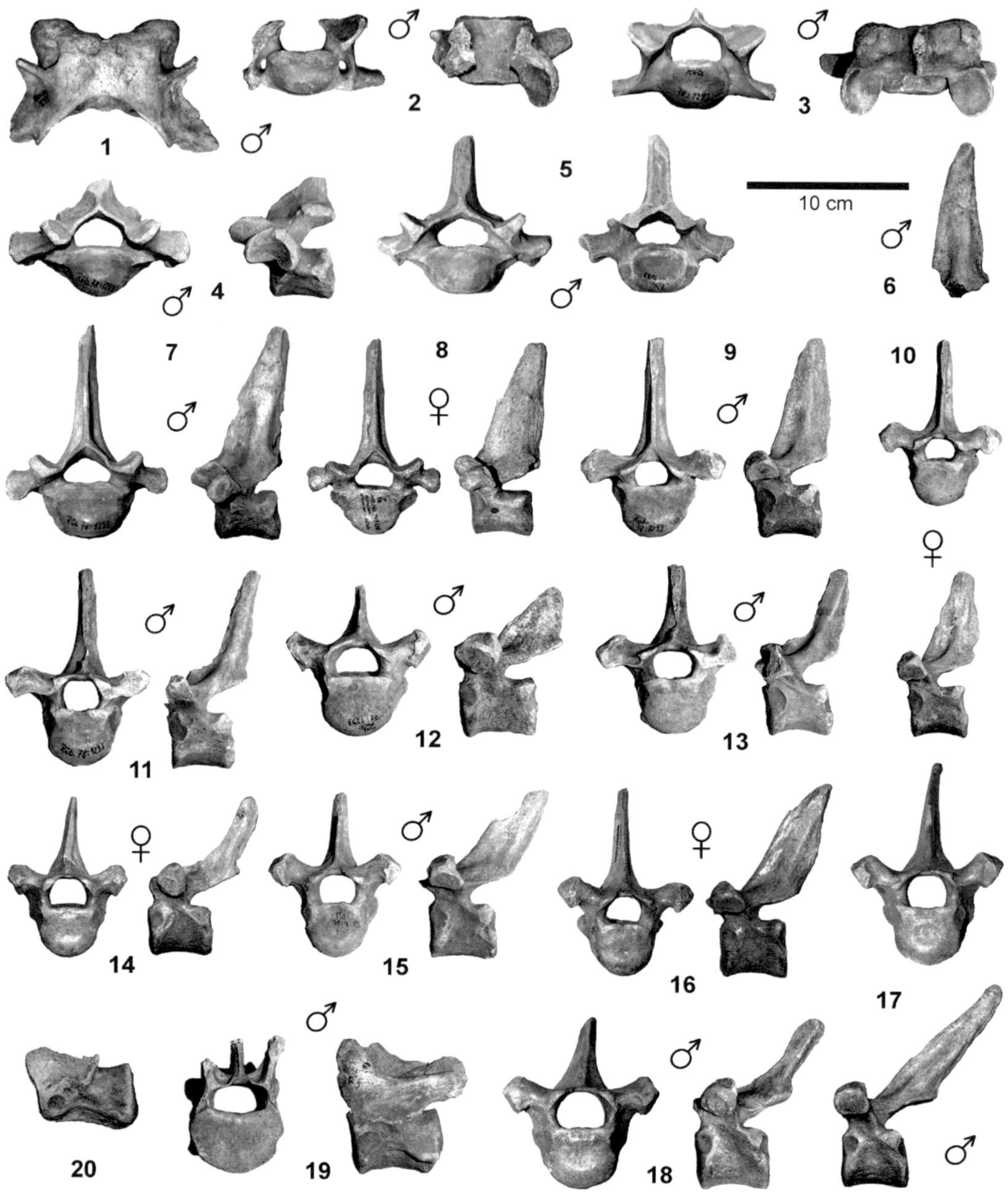

Figure 8. Axial skeleton bones (cervical to thoracic vertebrae) of *Panthera leo spelaea* (Goldfuss, 1810) var. *intermedia* nov. var. from the Hermann's Cave, Rübeland, Saxony-Anhalt (Germany). 1. Atlas (LDA no. 78:1293n), dorsal. 2. C5 (RCC no. HH-5a), cranial and dorsal. 3. C7 (LDA no. 78:1293o), cranial and dorsal. 4. T1 (LDA no. 78:1293p), cranial and dorsal. 5. T1 (LDA no. 78:1293q), cranial and dorsal. 6. T dorsal spine (MB no. Ma.14323), lateral. 7. T2 (LDA no. 78:1293r), cranial and lateral. 8. T2 (MB no. Ma.14305), cranial and lateral. 9. T3 (LDA no. 78:1293s), cranial and lateral. 10. T3 (LDA no. 78:1293t), cranial and lateral. 11. T4 (LDA no. 78:1293u), cranial and lateral. 12. T, middle (LDA no. 78:1293v), cranial and

lateral. 13. T, middle (LDA no. 78:1293w), cranial and lateral. 14. T5 (MB no. Ma.14308), cranial and lateral.15. T6 (LDA no. 78:1293x), cranial and lateral. 16. T6 (MB no. Ma.14307), cranial and lateral. 17. T7 (MB no. Ma.14304), cranial and lateral. 18. T8 (MB no. Ma.14306), cranial and lateral. 19. T13 (LDA no. 78:1293y), cranial and lateral. 20. T13-14 half centrum (RCC no. HH-5c), lateral.

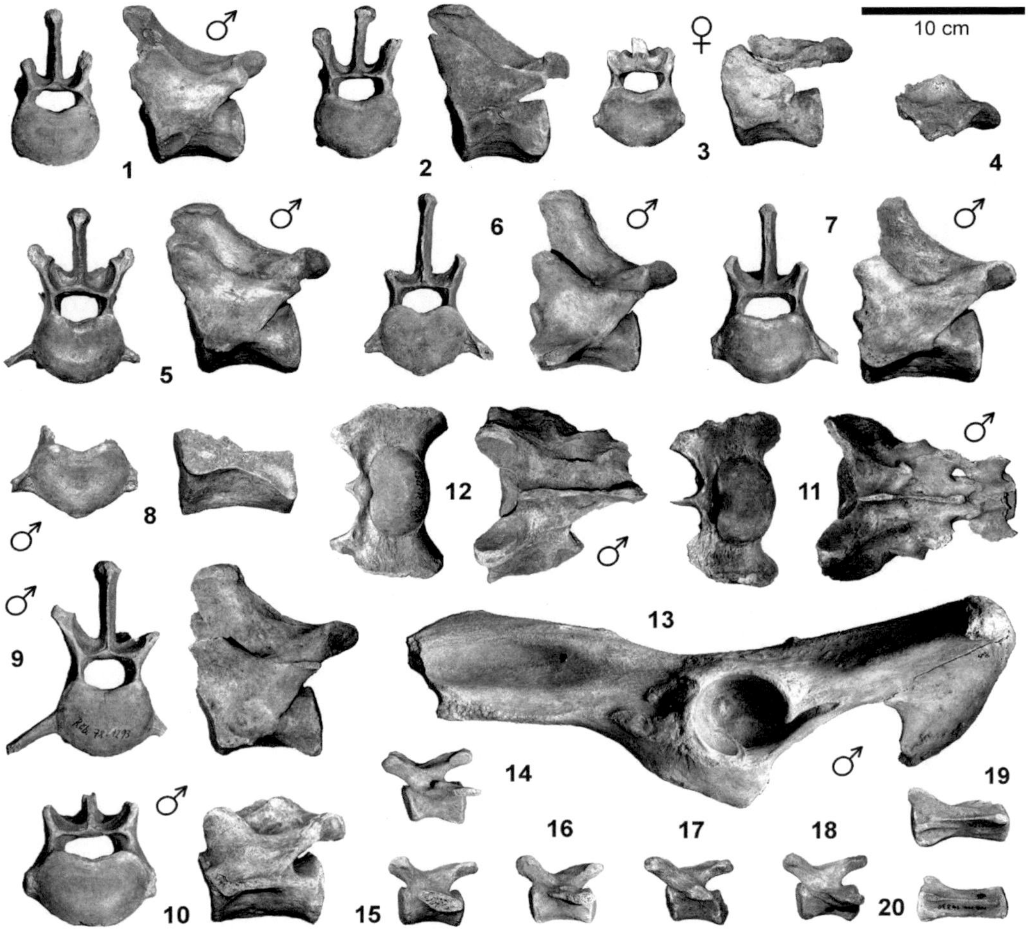

Figure 9. Axial skeleton bones (lumbar to caudal vertebrae, pelvis) of four Late Pleistocene (MIS 5a-d) *Panthera leo spelaea* (Goldfuss, 1810) var. *intermedia* nov. var. skeletons from the Hermann's Cave, Rübeland (Saxony-Anhalt). 1. T14 (LDA no. 78:1293z), cranial and lateral. 2. T14 (LDA no. 78:1293aa), cranial and lateral. 3. L1-2 (RCC no. HH-1), cranial and lateral. 4. L(?4) neural arch fragment (MB no. Ma. 14311b), lateral. 5. L1-2 (MB no. Ma. 14303), cranial and lateral. 6. L3-4 (LDA no. 78:1293bb), cranial and lateral. 7. L5 (MB no. Ma. 14302), cranial and lateral. 8. L(?4) centrum fragment (MB no. Ma. 14311a), cranial and lateral. 9. L5 (LDA no. 78:1293cc), cranial and lateral. 10. L7 (LDA no. 78:1293dd), cranial and lateral. 11. Sacrum (RCC no. HH-11), cranial and dorsal. 12. Sacrum (LDA no. 78:1293ee), cranial and dorsal. 13. Coxa (MB no. Ma. 14331), lateral. 14. Upper caudal vertebra (LDA no. 78:1293ff), lateral. 15. Upper caudal vertebra (LDA no. 78:1293gg), lateral. 16. Upper caudal vertebra (LDA no. 78:1293hh), lateral. 17. Upper caudal vertebra (MB no. Ma. 14310), lateral. 18. Middle caudal vertebra (MB no. Ma. 14309), lateral. 19. Middle caudal vertebra (MB no. Ma. 14330), lateral. 20. Lower caudal vertebra (LDA no. 78:1293ii), lateral (metrics see Table **1**, Fig. **12**).

stocene (MIS 9-3) steppe lion *P. l. spelaea* [2 - 5, 30]. The direct comparison and upper jaw P4 tooth metrics or lower jaw M1 data (Table **1**) fit into the higher evolved group of *P. l. spelaea.*

The new and updated bone metric statistics (crania and longbones) demonstrates well a size increase of the lions according to the Bergmann's rule (increase of body size in correlation to temperature drop) twice, in the late Middle (MIS 9-6) and then in the Late Pleistocene (MIS 5e-3). Within the Mid-Pleistocene Holsteinian Interglacial (MIS 9), small lions appeared in Central Europe from Africa with the hippo/water buffalo-fauna over the Donau and finally Rhine Rivers up to southern England [30]. From those small forms, which were also found at the Holsteinien Interglaical (MIS 9) archaeological Neanderthal hunter camp site Schöningen (only 50 km far northern from the Baumann's/Hermann's Cave in Rübeland) [40, 41].

In a first cycle developed very large forms in the glacials of the MIS 8 and 6. In a second cycle, the Eemian (MIS 5e) lions are again similar small as during the older MIS 21-9 Interglacial periods [40]. Those can be followed in the abundant Late Pleistocene Central European fossil record to increase in size with the temperature drop starting in the MIS 5a-d and increased to large forms (reching even nearly similar in size to the older *P. l. fossilis* Reichenau, 1906) in the cold MIS 4 period (Figs. **12, 13**) [40]. The last European lions, such as the skeleton of the large individual of Huttenheim, are from the Upper Rhine River terrace, dating cleary to the MIS 3 and early MIS2 before the Last Glacial Maximum (= LGM). Also one humerus from the Hermann's Cave falls into thos last large variants (Figs. **7.2, 12**). The Hutenheim and other lion remains from the Upper Terrace reach nearly similar large sizes as Cromerian Glacial stage lions of *P. l. fossilis* [cf. 25, 37 - 39] (large *P. l. spelaea* males seem to be in similar sizes of smaller *P . l. fossilis* females [40]). Lion size variants become now perfect for climate estimations and are very useful in biostratigraphy of the Late Pleistocene (Fig. **13**), whereas the Mid-Pleistocene material has to be analysed further. The *P. l. fossilis* from post-MIS 9 and even older Cromerian times are still confused and unsolved in both climatic and sexual dimorphism variant sizes.

Three late Mid- to Late Pleistocene variants are recently distinguished [40] with the small Holsteinian and Eemian Interglacial (MIS 9e, 5e) *P. l. s. brachygnathus* Dietrich, 1968 (holotype mandible from Dechen Cave [25], NW Germany and paratype skeleton of Neumark Nord Lake 1, Germany [40]) the most common Weichselian/Wuermian (MIS 4-5d) *P. l. s. intermedia* nov. var. (holotype skull and all figured paratype material from the Hermann's Cave, Figs. **6-11**), and the largest form *P. l. s. maximus* nov. var. (holotype skeleton from Huttenheim, SW Germany [36, 40]). Parallel, already in the early in the Cromerian (MIS 21), there

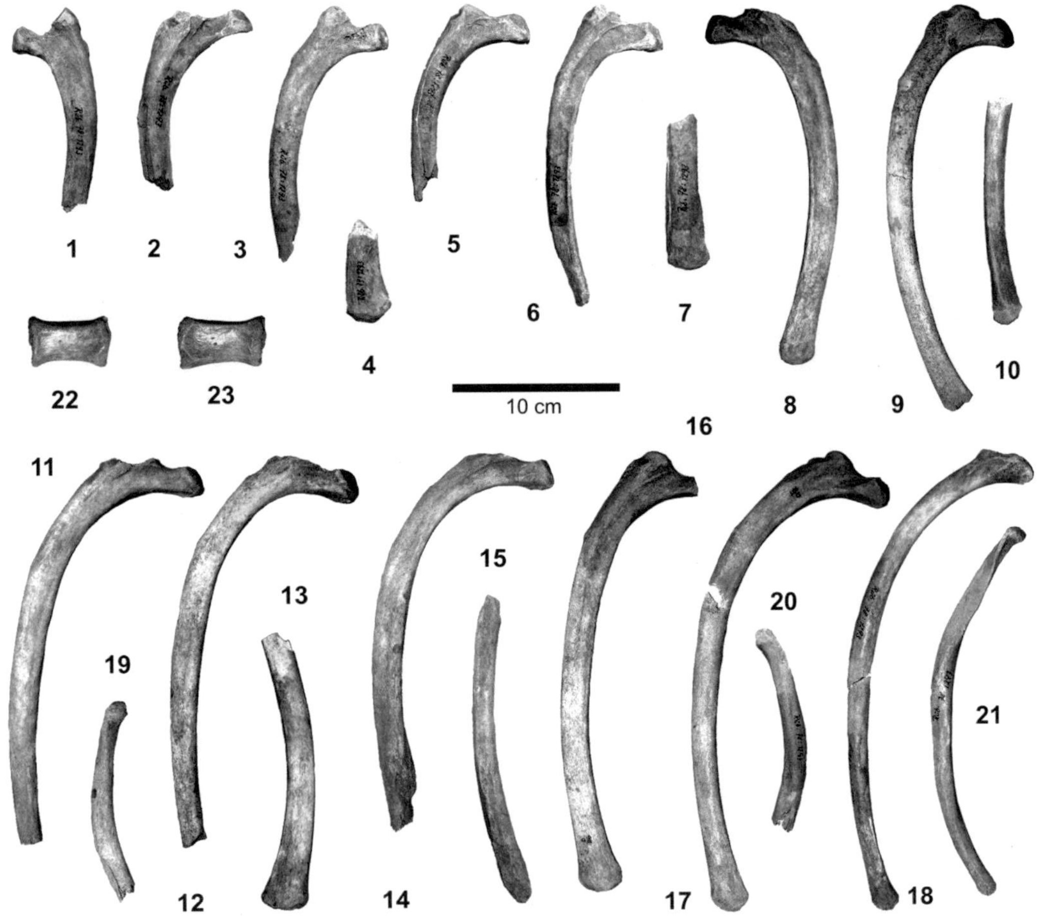

Figure 10. Axial skeleton bones (costae, sternal bones) of *Panthera leo spelaea* (Goldfuss, 1810) var. *intermedia* nov. var. from the Hermann's Cave, Rübeland, Saxony-Anhalt (Germany). 1. Left rib no. 1-2 (LDA no. 78:129jj). 2. Right rib no. ?4 (LDA no. 78:1293kk). 3. Right rib no. ?3 (LDA no. 78:1293ll). 4. Distal anterior rib fragment (LDA no. 78:1293mm). 5. Right rib no. ?5 (LDA no. 78:1293nn). 6. Right rib no. ?4 (LDA no. 78:1293oo). 7. Distal anterior rib fragment (LDA no. 78:1293pp). 8. Left rib no. 5 (MB no. Ma.14748). 9. Right rib no. 6 (MB no. Ma.51522). 10. Distal half middle rib (? No. 7) (LDA no. 78:1293qq). 11. Right middle proximal half rib (? No. 8) (LDA no. 78:1293rr). 12. Right middle proximal half rib (? No. 8) (LDA no. 78:1293ss). 13. Left middle distal half rib (RCC no. HH-3), lateral. 14. Right middle proximal half rib (?no. 9), with pathology (callus) (LDA no. 78:1293tt). 15. Middle distal rib fragment (LDA no. 78:1293uu). 16. Right rib no. 6 (MB no. Ma.51520). 17. Right rib no. 7 (MB no. Ma.51521). 18. Left rib no. 11 (LDA no. 78:1293vv). 19. Right rib no. 12-13 (LDA no. 78:1293ww). 20. Left rib no. 12-13 (LDA no. 78:1293xx). 21. Right rib no. 13 (LDA no. 78:1293yy). 22. Middle sternal bone (MB no. Ma. 11736). 23. Middle sternal bone (MB no. Ma. 11734). All lateral.

appears a small interglacial lion form (same sizes as MIS 9, 5e), whereas *P. l. fossilis* as the "largest lion" of Europe must be limited after the Bergmann's rule results for the Late Pleistocene lions [40, Tables **2-7**] on the glacial periods. If so,

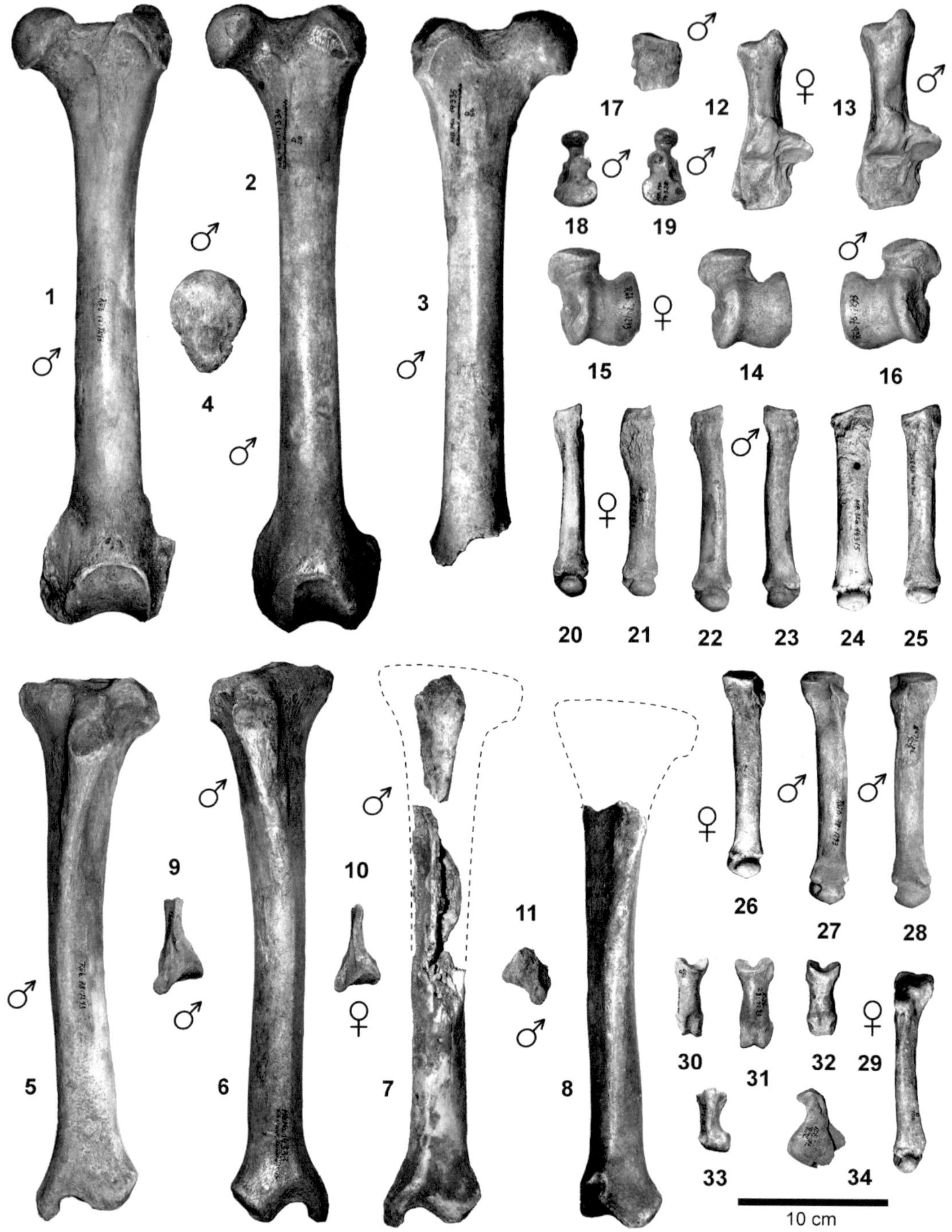

Figure 11. Hind limb bones of *Panthera leo spelaea* (Goldfuss, 1810) var. *intermedia* nov. var. from the Hermann;s Cave, Rübeland, Saxony-Anhalt (Germany). 1. Left femur (LDA no. 78:1293xc), cranial. 2. Left

femur (MB no. Ma. 14334), cranial. 3. Right femur (MB no. Ma. 14335), cranial. 4. Patella (MB no. Ma. 9541), cranial. 5. Right tibia (LDA no. 78:1293xd), cranial. 6. Left tibia (MB no. Ma. 14333), cranial. 7. Right tibia (RCC no. HH-9), cranial. 8. Right tibia from a small subadult female (RCC no. HH-2), cranial. 9. Right distal fibula joint (LDA no. 78:1293xe), lateral. 10. Right distal fibula joint (LDA no. 78:1293xf), lateral. 11. Left distal fibula joint (MB no. Ma. 14324), lateral. 12. Right calcaneus (LDA no. 78:1293xg), cranial. 13. Right calcaneus (LDA no. 78:1293xh), cranial. 14. Left astragalus (LDA no. 78:1293xi), dorsal. 15. Left astragalus (LDA no. 78:1293xj), dorsal. 16. Right astragalus (LDA no. 78:1293xk), dorsal. 17. Left cuboid (LDA no. 78:1293xl), cranial. 18. Left intermedium (MB no. Ma. 14327), dorsal. 19. Right intermedium (MB no. Ma. 14328), dorsal. 20. Left Mt II (MB no. Ma. 14312), dorsal. 21. Left Mt II, pathology (exostosis) (LDA no. 78:1293xm), dorsal. 22. Left Mt II, (LDA no. 78:1293xn), dorsal. 23. Right Mt II (LDA no. 78:1293xo), dorsal. 24. Left Mt III (MB no. Ma. 14315), dorsal. 25. Left Mt III (MB no. Ma. 14313), dorsal. 26. Right Mt IV from a small subadult female (MB no. Ma. 14316), dorsal. 27. Right Mt IV (LDA no. 78:1293xp), dorsal. 28. Left Mt IV (LDA no. 78:1293xq), dorsal. 29. Left Mt V from a small subadult female (MB no. Ma. 14314), dorsal. 30. Phalanx I, ?digit V (MB no. Ma. 9869), dorsal. 31. Phalanx I, right ?digit II (LDA no. 78:1293xa), dorsal. 32. Phalanx I (MB no. Ma. 14712), dorsal. 33. Phalanx II, ?right digit IV (LDA no. 78:1293xr), dorsal. 34. Phalanx III (LDA no. 78:1293xb), dorsal (metrics see Table **1**, Fig. **12**).

there is another third similar Bergmann's rule trend already in the early Middle Pleistocene starting with small MIS 21 lions, which devellop in the largest lion forms at all somehow in Cromerian Glacial stages [40]. With this new subdevision, the sexual dimorphism was partly corrected [40], which was confused in former times with glacial/interglacial variants in many cases.

Pathologies

All three preserved elderly pathologic female skull canines (both upper, and the lower left) are flaked by use and are further polished (Figs. **6**, **14**) and those tip-less teeth had no canine bite force or penetration function anymore. In addition no full function remained of the left scissor dentation because of the M1 cusp flaking damage (Figs. **6**, **14**). All other teeth have stronger wear which signals rubbing of the cusps. The massive tooth wear tip damages in comparison to modern African lions are quite unusual because extant lions mainly consume cartilage and soft meat, cartilage or bone spongiosa and their teeth remain mostly unflaked, broken or damaged up to elderly ages [42 - 45]. The wear damage of the elderly lion has similarities to their antagonists, the Ice Age spotted hyenas, which also demonstrate in older individual ages stronger tooth use and flaking which resulted from long-term and massive bone cutting and breaking [46]. The dentition of the elderly lion reflect a higher bone consumption due to cutting or cracking of smaller or even large bone parts of more unusual prey. This sample correlates to the cave bear feeding specialization of lions which is known for Late Pleistocene European mountainous regions [4, 48]. Whereas dental damages appear only in very old individuals and rarer on the canines such as at the female from the Hermann's Cave (Figs. **6**, **14**), the maximum wear stages of European steppe lions shows only tip worn dentition [49].

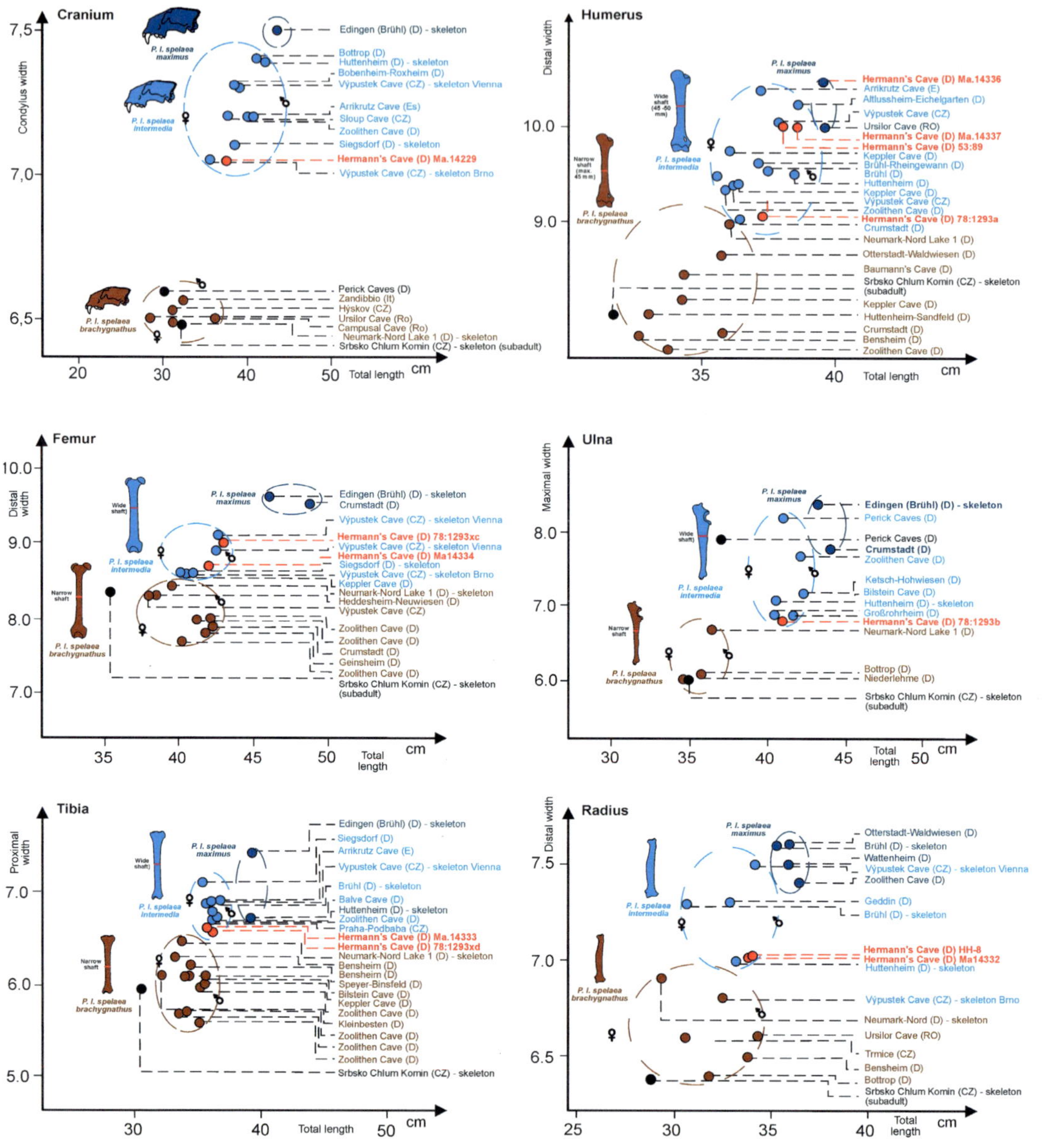

Figure 12. Osteometric comparison of the four Hermann's Cave steppe lions to other European sites (comparative data based on [4, 13 - 15, 30 - 34, 36, 40] modified from [40]). The black dots (Srbsko Chlum Komin Cave skeleton, Perick Caves) fall out of the range circles of their species due to their subadult individual ages, which represent not full grown up animals that confuse plots. Blue: Glacial lions, Brown: Interglacial lions. Red: Hermann's cave material.

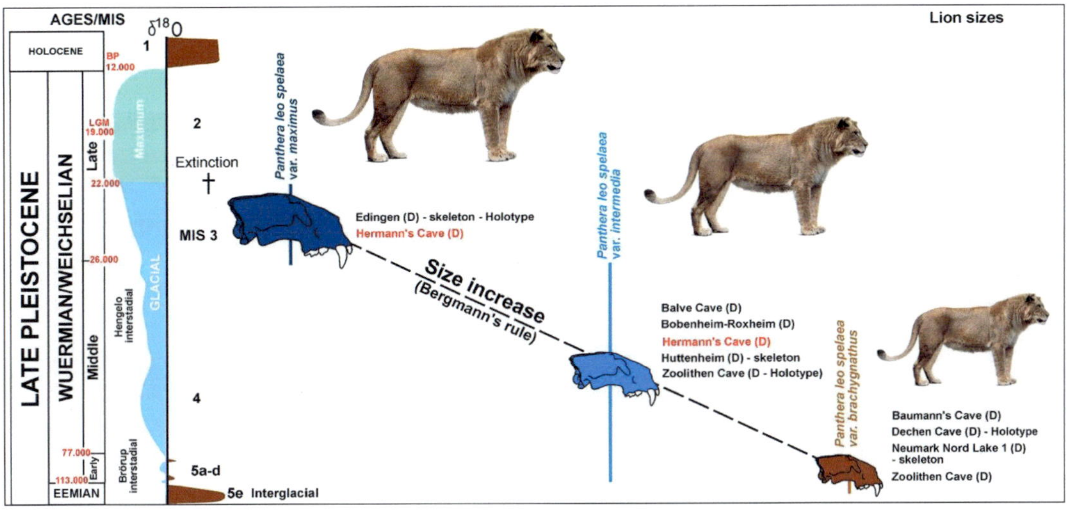

Figure 13. Model of increase of the lion size within the late Pleistocene of Central Europe with importanrt sites in Germany (data see Tables **2-7**) with lion sizes of the Hermann's Cave. Those date into the MIS 5d-3, which correlate to the cave bear clock dating (Fig. **5**).

Two different types of trauma pathogies are present on the material. The first is a rib fracture, which was in the healing process and with callus thickening in the middle part of a posterior rib (Fig. **14**). Another unhealed trauma resulted in exostoses on the dorsal surface of the left metatarsus III. Both trauma could have resulted from both, inter- or intraspecies fights [cf. 4, 46, 49]. Pedal bite trauma, such as in this case, seem to result most probable from battles with hyenas [46, 49]. For extinct and extant spotted hyenas described, those are attacking their antagonists and prey from behind biting the enemy into the hind leg or hind foot [49, 50]. Extant lions attack opposite from the front using their paw claws and their canines for the struggle [*e.g.* 42, 49]. For the elderly pathogenic lion from the Sall Hall a most probable death explanation possibility can be reconstructed. The lion was obviously received a mortal wound in a battle conflict with a hyena clan or with other lion's most probably outside the cave, but he did not dwell in the cave after the injury. He must have entered about one to two weeks later when the bone trauma injuries were already in a healing process. Either he died there as reason of his injuries (*e.g.* final bacterial inflammation, no food and too weak), or he died as very weak attacker in a lost battle with a cave bear, who would be protecting her cubs during winter hibernation. Those would have been easy kills for lions during winter times, even for a sickly lion. If the cub was protected well, the lion lost its final battle with an adult cave bear. The lion carcass was left untouched after its death on place in the Saal Hall cave bear hibernation area, because the large full herbivorous late cave bear species *Ursus spelaeus ingressus* did not consume meat [51], and hyenas did not dwell generally that deep into caves of cave bear dens [8].

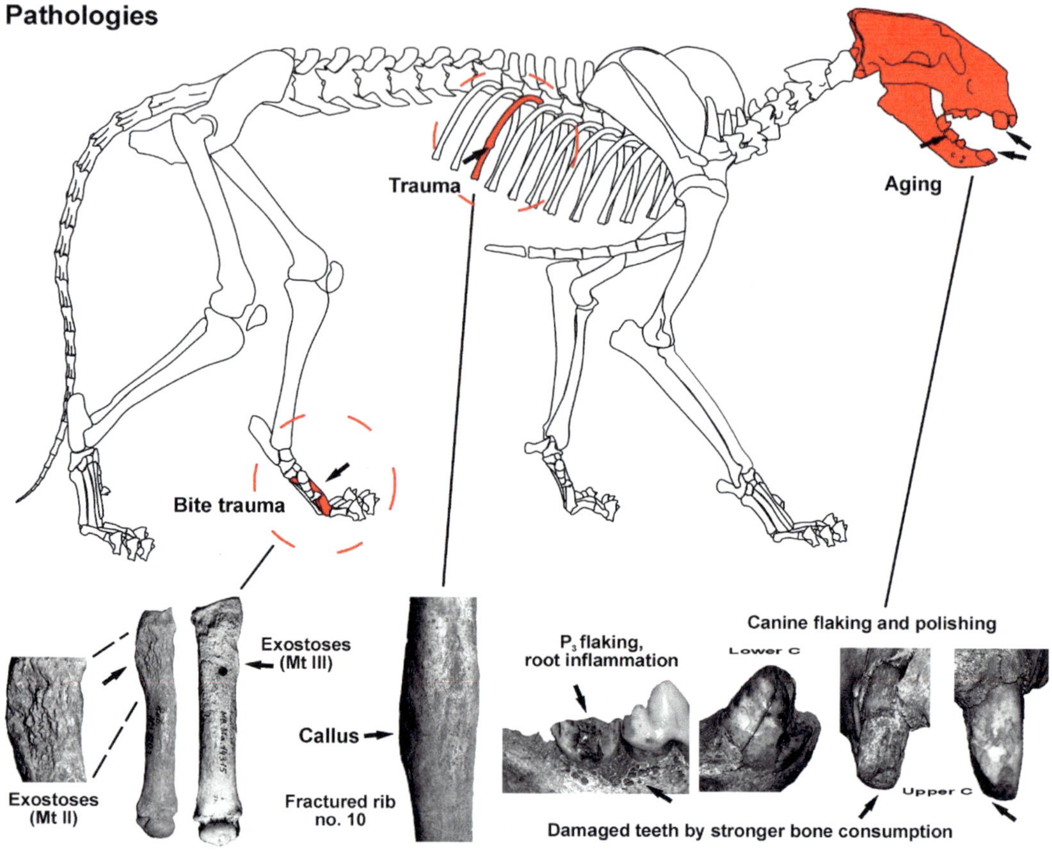

Figure 14. Dental and postcranial pathologies of the *Panthera leo spelaea* (Goldfuss, 1810) var. *intermedia* nov. var. skeleton possibly all belonging to a single elderly female individual from the Sall Hall of the Herman's Cave, Rübeland, Saxony-Anhalt, Germany.

Steppe Lions as Specialized Cave Bear Hunters

It is of importance for the understanding of the breeding and hunting strategy of the last Ice Age steppe lions of Europe is their individual age and sex ratio calculated on material which is found either in caves, or at open air sites [cf. 8]. In contrast to the Late Pleistocene spotted hyenas, the prey of the lions is much more difficult to reconstruct, because they did not accumulate prey or bone remains of their prey or do not leave consistently damaged bones [*e.g.* 42 - 43]. The theory about lions killing cave bears even in deep and large cave systems resulted from intensive taphonomic cave studies [3, 4, 8, 28, 31, 33, 34, 46], whereas convincing evidence has recently emerged from Urşilor Cave cave bear den

(Ro, Fig. **4**) [48, 52]. Further evidence from taphonomic studies is demonstrated from several caves in the Sauerland Karst [8]. Similar results were found at the Zoolithen Cave with the largest steppe lion population found ever in a large cave bear den (about a half million cave bear bones), where a feeding specialization on cave bears was argued due to the absence of mammoth steppe fauna in mountainous regions [52]. The latest nitrogen isotope data supports the cave bear hunting interpretation, whereas other researchers went further to declare that only "cubs" were a main food source [53]. Recent taphonomic studies at the Sophie's Cave support this [54]. The hunting of cave bears by lions explains why cave bears hibernated as deep as possible in larger cave systems even in dangerous passages to protect themselves and the cubs against lion attacks [8, 47]. The aforementioned "normal mortality" in cave bears of Europe described for several caves in Europe [cf. *e.g.* 55 - 57] does not exist. The high mortality of cave bear cubs (also Hermann's Cave, see Chapter Cave bears) is a combination of both: natural dying (especially neonate aborts up to 12 months young cubs) and predation mortality, which is demonstrated by the percentage of bite damaged cub bones, whereas most of the bite damages clearly are of postmortal times (*e.g.* chewed joints, see Chapter Cave bears). Also the picture of the cave bears life and death has been presented incorrectly in the past, such as in the bone taphonomy: cave bears as "cannibals" [58, 59]), by not taking the predators into any account in any cave bear population statistics/mortality analyses. Because of the lack of modern lion/bear population overlap in Africa, there are no possibilities to compare this very unique situation of lions hunting cave bears deep in caves during the Late Pleistocene in Europe.

Figure 15. Comparison of male steppe lion *Panthera leo spelaea* var. *brachygnathus/maximus* and its cave bear guilt: (also males) the smaller cave bear *Ursus spelaeus ingressus/spelaeus/eremus* (Illustration © R. Uchytel).

Lion or Lioness – Single or Pack Cave Bear Killers?

The most interesting question about the hunting strategies of the last lions of Europe has not been resolved, and difficult to prove anyway. As known for extant African lions whose hunting strategies differ either as single, in two or in female packs depending mostly on the guilt type. Furthermore, they also steal carcasses from other predators, such as from hyenas [*e.g.* 42 - 43]. Hunting cave bears in the caves seems to have performed mainly by females, whereas this might be different in each cave due to different reasons. There are some larger cave bear dens in Europe (*e.g.* Keppler Cave, Urşilor Cave), where two third of the lion bone material belongs to females [33, 48, 52]. The male/female ratio at mixed hyena/cave bear den sites is opposite (Bilstein Cave, Perick Caves [33], Zoolithen Cave [4]) with only one third of the bones from female hyenas.

In the new case of the Hermann's Cave the ratio correlates to the latter cases. There are three males and one female (3-1 ratio), all representing grown up to elderly individuals. Very typical for nearly all caves in Central Europe (cave bear dens) is the absence of lion cub material [8]. This is represented only as guilt in caves which were used as hyena dens (lion remains in hyena bone assemblages [8, 29]). Already for other studied caves this absence excludes the cave use as lion den [8]. Lion cubs must have been raised, similar as in modern African lions, only in the Ice Age open mammoth steppe or boreal forest palaeoenvironments [8]. The Hermann's Cave material also supports this theory with the absence of cub or subadult skeletal remains. This additionally leads to the reconstruction of the lion cub-subadult 3D modelling with a protective fur colour and ornamentation (cf. Fig. **3**). Lion cubs or subadults never entered cave bear dens and were obviously no cave bear killers in that early age [8].

As found similar in four other convincing examples (Fig. **16**), only adult to elderly and hunting specialist lions attempted the cave bear hunt into larger cave bear cave systems [8, 28, 48]. In the case of the Hermann's Cave, more males as females have been killed inside the cave (3 males/1 female - ratio). The amount cannot be used uncritically to argue only for a cave bear hunt that was mainly performed by males. Possibly, males were simply the less successful cave bear hunters and failed more often, or they hunted cave bears more often. Again, compared to African lions, the males are the attackers and killers of other carnivores, especially other lions or hyenas [51, 58], which is possibly also reflected in their trauma pathologies. They also hunt as single or with two larger or more dangerous guilt, whereas the females hunt most successful in larger packs herbivorous guilt [42, 43]. If Ice Age lions hunted the same or similar ways, the cave bear hunters must have been mainly males which were able to fight or kill adult cave bears, whether the smaller or the larger cave bears. In the Late Pleistocene in Central Europe there were in non-alpine regions the smaller *Ursus spelaeus eremus* and the largest cave bear *Ursus spelaeus ingressus* (Fig. **15**), which both are also well recorded from the Hermann's Cave (see Chapter cave bears). The smaller cave bears could have been killed by a single lioness, which

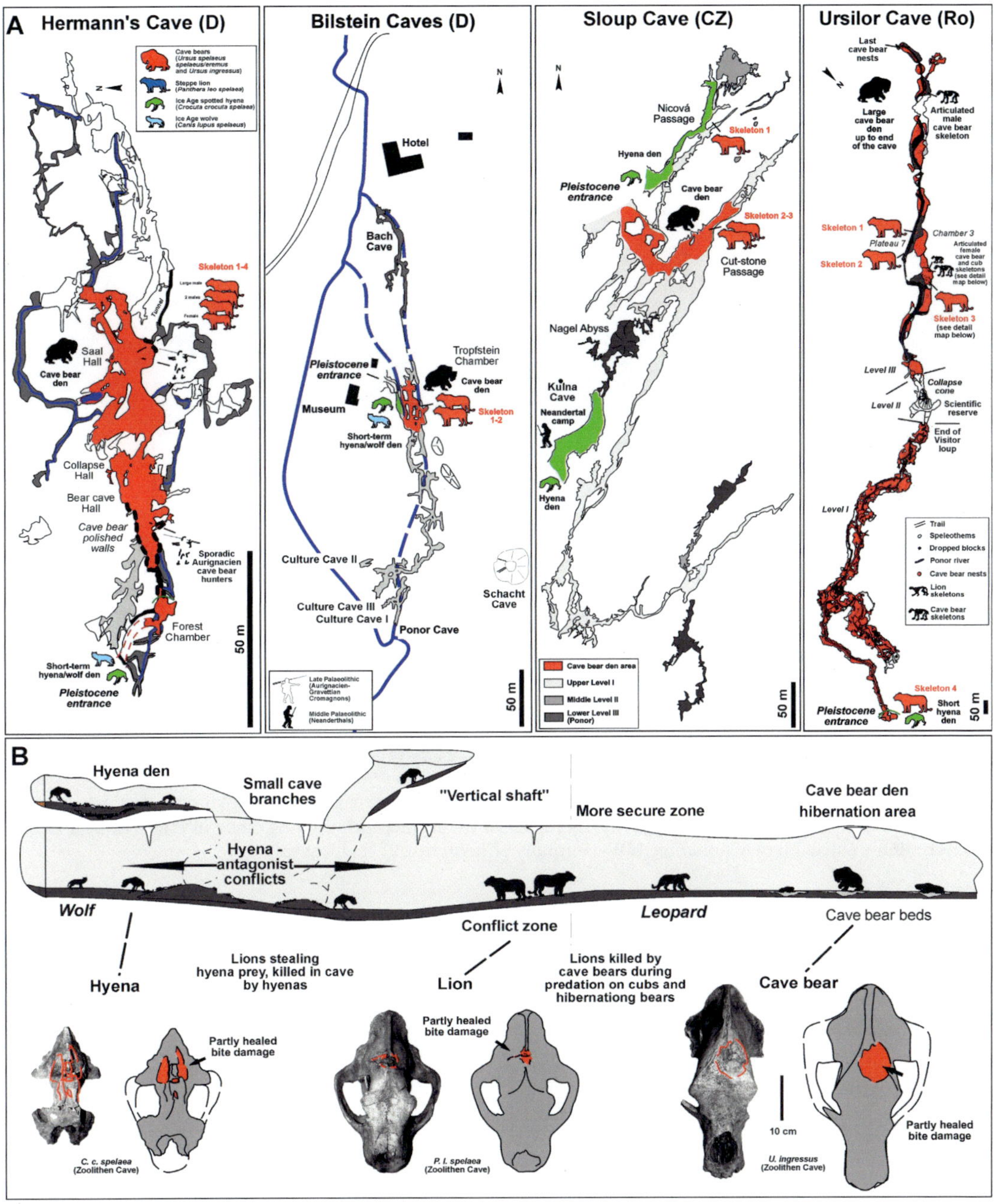

Figure 16. A. European cave bear den caves with prove of cave bear predation of grown up lions which are often found as articulated or scattered skeletons deeper in the caves. Added herein A. Hermann's Cave; composed and modified: B. Bilstein Cave, modified from [28], C. Sloup Cave, modified from [34], D-E. Best example at Urşilor Cave, where three articulated lion skeletons have been found deep in the cave on the cave bear hibernation plateaus, one completely articulated, and one scattered as result of scavenging – possibly lion cannibalism, modified from [48, 52].

was of similar size, but lighter in weight (Fig. **15**). There is no proof yet to speculate whether lionesses killed preferably smaller cave bears, and males the larger cave bear species.

Figure 17. A large male cave bear *Ursus spelaeus ingressus* kills a weak elderly dental pathological male Ice Age steppe lion *Panthera leo spelaea* in the Saal Hall 90 m distance from the entrance in the cave bear den of the Hermann's Cave, Harz Mountains, NW-Germany (Illustration © R. Uchytel).

The Lost Battle Deep in Cave Bear Dens

Recently three convincing examples of lion skeleton records are published which seem to be the result of lost fights: a. Sloup Cave "skeletons" (at minimum 2) [34], b. Urşilor Cave skeletons (at minimum 3 [48],), c. Zoolithen Cave (unclear amount, but remains are from originally partially articulated skeletons [4]). In at least the elderly pathogenic individual from the Hermann's Cave it is best to argue, why this very old individual lost the fight with by a cave bear. All four lion canines were flaked (Fig. **13**), round polished and useless for any kill-bite.

Even if the attack was made mainly by the paw claws, the postcranial skeleton shows further pathologies and strong weakness of the lion (Fig. **13**).

To those, herein another cave can be added with one of the best convincing examples (similar as Urşilor Cave), the Hermann's Cave, which is the most northern European record. The Late Pleistocene steppe lion, compared with its body weight estimation to those of its cave bear guilt makes clear, that in a fight between only a single lion those must have been mainly successful even in killing the smaller cave bears *U. s. eremus* (Fig. **15**). Large *U. s. ingressus* cave bear must be expected to have been killed, for sure was not easy because of its greater weight and large canine weapons. The lion remains in bone-rich cave bear dens never exceed 1-3% off the total NISP bone amount [8]. This amount which is at the Hermann's Cave also only 2.2% calculates approximately to only one dead lion in a cave bear den cave due to lost battles (Fig. **17**) per 10,000 years at maximum, which suggests mainly successful hunting on the "hibernating" and easy to kill cubs over 100.000 years.

Table 1. Bone list of 128 of Late Pleistocene (MIS 5a-d) *Panthera leo spelaea* (Goldfuss, 1810) from four individual remains of the Hermann' s Cave in Rübeland, Saxony-Anhalt, Central Germany.

No.	Inv.-No.	Bone type	Commentary	Left	Right	Ind. Age	Sex		Old Collection	Collection
1	Ma.14299	Cranium	Incomplete, with rigth P3, C, and left C, P3-4 P4 width = 37 mm, length = 374 mm, condyl width = 75 mm			Elderly	Male		Excavation Völker 1984/85 (Quadrant D 137)	Museum für Naturkunde Berlin
2	Ma.14298	Mandible, belongs to cranium	Incomplete, with C, P3-M1 root	x		Elderly	Male		Excavation Völker 1984/85 (Quadrant D 62)	Museum für Naturkunde Berlin
3	Ma.14301	Scapula	Incomplete, glenoid width = 61 mm	x		Adult-Elderly	Male		Excavation Völker 1984/85 (Quadrant D 25)	Museum für Naturkunde Berlin
4	53:89	Humerus	Complete, length = 386 mm, distal width = 100 mm		x	Adult-Elderly	Male		Steiner 1968 Excavation	Landesmuseum Sachsen-Anhalt
5	78:1293a	Humerus	Complete, length = 376 mm, distal width = 94 mm	x		Adult-Elderly		Female	Steiner 1969 Excavation	Landesmuseum Sachsen-Anhalt

(Table 1) contd.....

No.	Inv.-No.	Bone type	Commentary	Left	Right	Ind. Age	Sex		Old Collection	Collection
6	Ma.14337	Humerus	Complete, length = 381 mm, distal width = 100 mm		x	Adult-Elderly	Male		Excavation Völker 1984/85 (Quadrant D 137)	Museum für Naturkunde Berlin
7	Ma.14336	Humerus	Complete, length = 396 mm, distal width = 104 mm		x	Adult-Elderly	Male		Excavation Völker 1984/85 (Quadrant D 94)	Museum für Naturkunde Berlin
8	78:1293b	Ulna	Nearly complete, modern damaged, length = 409 mm, largest width = 68 mm		x	Adult-Elderly	Male		Steiner 1968 Excavation	Landesmuseum Sachsen-Anhalt
9	Ma.14332	Radius	Complete, length = 338 mm, distal width = 70 mm	x		Adult-Elderly	Male		Excavation Völker 1984/85 (Quadrant F)	Museum für Naturkunde Berlin
10	HH-8	Radius	Complete, length = 339 mm, distal width 70 mm		x	Adult-Elderly	Male		Prospections Diedrich, 2012	Rübeland Cave Center
11	Ma.14322	Scapholunatum	Complete, with = 57 mm, length = 43 mm	x		Adult-Elderly	Male		Excavation Steiner Halle/Saale 1962 (81:1225)	Museum für Naturkunde Berlin
12	Ma.14326	Carpale 3	Complete		x	Adult-Elderly	Male		Excavation Völker 1984/85 (Quadrant D 149)	Museum für Naturkunde Berlin
13	Ma.14329	Pisiforme	Complete		x	Adult-Elderly	Male		Excavation Völker 1984/85 (Quadrant D 128)	Museum für Naturkunde Berlin
14	Ma.14321	Metacarpus	I, length = 48 mm, distal width = 20 mm	x		Adult-Elderly	Male		Excavation Völker 1984/85 (Quadrant D F176)	Museum für Naturkunde Berlin
15	Ma.14317	Metacarpus	II, length = 115 mm, distal width = 25 mm	x		Adult-Elderly		Female	Excavation Steiner Halle/Saale 1962 (81:1225)	Museum für Naturkunde Berlin

(Table 1) contd.....

No.	Inv.-No.	Bone type	Commentary	Left	Right	Ind. Age	Sex		Old Collection	Collection
16	78:1293c	Metacarpus	III, length 128 mm, distal width 27 mm, pathology	x		Adult-Elderly	Male		Steiner 1968 Excavation	Landesmuseum Sachsen-Anhalt
17	78:1293d	Metacarpus	III, length = 131 mm, distal width = 28 mm	x		Adult-Elderly	Male		Steiner 1968 Excavation	Landesmuseum Sachsen-Anhalt
18	78:1293e	Metacarpus	III, length = 131 mm, distal width = 28 mm		x	Adult-Elderly	Male		Steiner 1968 Excavation	Landesmuseum Sachsen-Anhalt
19	Ma.14320	Metacarpus	IV, length = 120 mm, distal width = 23 mm	x		Adult-Elderly	Male		Excavation Völker 1984/85 (Quadrant F 179)	Museum für Naturkunde Berlin
20	78:1293f	Metacarpus	IV, length = 128 mm, distal width = 23 mm	x		Adult-Elderly	Male		Steiner 1968 Excavation	Landesmuseum Sachsen-Anhalt
21	78:1293g	Metacarpus	IV, length 129 mm, Distal width 28 mm		x	Adult-Elderly	Male		Steiner 1968 Excavation	Landesmuseum Sachsen-Anhalt
22	78:1293h	Metacarpus	IV, length = 128 mm, distal width = 27 mm		x	Adult-Elderly	Male		Steiner 1968 Excavation	Landesmuseum Sachsen-Anhalt
23	78:1293i	Metacarpus	V, Length = 110 mm, distal width = 26 mm	x		Adult-Elderly	Male		Steiner 1968 Excavation	Landesmuseum Sachsen-Anhalt
24	Ma.14318	Metacarpus	V, length 104 mm, distal width 24 mm		x	Adult-Elderly		Female	Excavation Steiner Halle/Saale 1962 (81:1225)	Goldfussmuseum Bonn
25	Ma.14319	Metacarpus	V, Length = 104 mm, distal width = 24 mm		x	Adult-Elderly		?Female	Excavation Steiner Halle/Saale 1962 (81:1225)	Museum für Naturkunde Berlin

(Table 1) contd.....

No.	Inv.-No.	Bone type	Commentary	Left	Right	Ind. Age	Sex		Old Collection	Collection
26	Ma.9953	Phalanx I	Complete, digit V		x	Adult-Elderly			Excavation Völker 1984/85 (Quadrant D 159)	Museum für Naturkunde Berlin
27	Ma.9910	Phalanx I	Complete, ?digit IV		x	Adult-Elderly			Excavation Völker 1984/85 (Quadrant D 180)	Museum für Naturkunde Berlin
28	78:1293zz	Phalanx I	Complete, ?digit IV		x	Adult-Elderly			Steiner 1968 Excavation	Landesmuseum Sachsen-Anhalt
29	78:1293j	Phalanx I	Complete, ?digit III		x	Adult-Elderly			Steiner 1968 Excavation	Landesmuseum Sachsen-Anhalt
30	78:1293k	Phalanx I	Complete, ?digit III		x	Adult-Elderly			Steiner 1968 Excavation	Landesmuseum Sachsen-Anhalt
31	78:1293l	Phalanx II	Complete			Adult-Elderly			Steiner 1968 Excavation	Landesmuseum Sachsen-Anhalt
32	78:1293m	Phalanx III	Complete			Adult-Elderly			Steiner 1968 Excavation	Landesmuseum Sachsen-Anhalt
33	78:1293n	Cervical vertebra	Atlas, incomplete			Adult-Elderly	Male		Steiner 1968 Excavation	Landesmuseum Sachsen-Anhalt
34	HH-5a	Cervical vertebra	C5			Adult-Elderly	Male		Prospections Diedrich, 2012	Rübeland Cave Center
35	78:1293o	Cervical vertebra	C7			Adult-Elderly	Male		Steiner 1968 Excavation	Landesmuseum Sachsen-Anhalt
36	78:1293p	Thoracic vertebra	T1, incomplete			Adult-Elderly	Male		Steiner 1968 Excavation	Landesmuseum Sachsen-Anhalt
37	78:1293q	Thoracic vertebra	T1, incomplete			Adult-Elderly	Male		Steiner 1968 Excavation	Landesmuseum Sachsen-Anhalt
38	Ma.14323	Thoracic vertebra	T1-2, proc spinosus			Adult-Elderly	Male		Excavation Steiner Halle/Saale 1962 (81:1225)	Museum für Naturkunde Berlin
39	78:1293r	Thoracic vertebra	T2, incomplete			Adult-Elderly	Male		Steiner 1968 Excavation	Landesmuseum Sachsen-Anhalt
40	Ma.14305	Thoracic vertebra	T2, incomplete			Adult-Elderly		Female	Excavation Völker 1984/85 (Quadrant F 78)	Museum für Naturkunde Berlin
41	78:1293s	Thoracic vertebra	T3, incomplete			Adult-Elderly	Male		Steiner 1968 Excavation	Landesmuseum Sachsen-Anhalt
42	78:1293t	Thoracic vertebra	T3, incomplete			Adult-Elderly	Male	Female	Steiner 1968 Excavation	Landesmuseum Sachsen-Anhalt
43	78:1293u	Thoracic vertebra	T4, incomplete			Adult-Elderly	Male		Steiner 1968 Excavation	Landesmuseum Sachsen-Anhalt

(Table 1) contd.....

No.	Inv.-No.	Bone type	Commentary	Left	Right	Ind. Age	Sex		Old Collection	Collection
44	78:1293v	Thoracic vertebra	T - middle, incomplete			Adult-Elderly	Male		Steiner 1968 Excavation	Landesmuseum Sachsen-Anhalt
45	78:1293w	Thoracic vertebra	T - middle, incomplete			Adult-Elderly	Male		Steiner 1968 Excavation	Landesmuseum Sachsen-Anhalt
46	Ma.14308	Thoracic vertebra	T5, incomplete			Adult-Elderly		Female	Excavation Völker 1984/85 (Quadrant F 181)	Museum für Naturkunde Berlin
47	78:1293x	Thoracic vertebra	T6, incomplete			Adult-Elderly	Male		Steiner 1968 Excavation	Landesmuseum Sachsen-Anhalt
48	Ma.14307	Thoracic vertebra	T4-5, incomplete			Adult-Elderly		Female	Excavation Völker 1984/85 (Quadrant F 45)	Museum für Naturkunde Berlin
49	Ma.14304	Thoracic vertebra	T7, complete			Adult-Elderly	Male		Excavation Völker 1984/85 (Quadrant D 111)	Museum für Naturkunde Berlin
50	Ma.14306	Thoracic vertebra	T8, incomplete			Adult-Elderly	Male		Excavation Völker 1984/85 (Quadrant D 90)	Museum für Naturkunde Berlin
51	HH-5c	Thoracic vertebra	?T13-14, half centrum			Adult-Elderly	Male		Prospections Diedrich, 2012	Rübeland Cave Center
52	78:1293y	Thoracic vertebra	T13, incomplete			Adult-Elderly	Male		Steiner 1968 Excavation	Landesmuseum Sachsen-Anhalt
53	78:1293z	Thoracic vertebra	T14, incomplete			Adult-Elderly	Male		Steiner 1968 Excavation	Landesmuseum Sachsen-Anhalt
54	78:1293aa	Thoracic vertebra	T14, incomplete			Adult-Elderly	Male		Steiner 1968 Excavation	Landesmuseum Sachsen-Anhalt
55	HH-1	Lumbar vertebra	L1-2, incomplete			Adult-Elderly		Female	Prospections Diedrich, 2012	Rübeland Cave Center
56	Ma.14303	Lumbar vertebra	L1-2			Adult-Elderly	Male		Excavation Völker 1984/85 (Quadrant D 128)	Museum für Naturkunde Berlin
57	78:1293bb	Lumbar vertebra	L3-4, incomplete			Adult-Elderly	Male		Prospections Diedrich, 2012	Rübeland Cave Center
58	Ma.14311 (a and b)	Lumbar vertebra	Possibly L4 centrum fragment and neural arch fragment			Adult-Elderly	Male		Excavation Völker 1984/85 (Quadrant D167/189)	Museum für Naturkunde Berlin

(Table 1) contd.....

No.	Inv.-No.	Bone type	Commentary	Left	Right	Ind. Age	Sex		Old Collection	Collection
59	Ma.14302	Lumbar vertebra	L5			Adult-Elderly	Male		Excavation Steiner Halle/Saale 1962 (81:1225)	Museum für Naturkunde Berlin
60	78:1293cc	Lumbar vertebra	L5			Adult-Elderly	Male		Steiner 1968 Excavation	Landesmuseum Sachsen-Anhalt
61	78:1293dd	Lumbar vertebra	L7			Adult-Elderly	Male		Steiner 1968 Excavation	Landesmuseum Sachsen-Anhalt
62	Ma.14331	Pelvis	Coxa, acetabulum width = 52 mm	x		Adult	Male		Excavation Völker 1984/85 (Quadrant D 109/61)	Museum für Naturkunde Berlin
63	HH-10	Pelvis	Sacrum			Adult	Male		Prospections Diedrich, 2012	Rübeland Cave Center
64	78:1293ee	Pelvis	Sacrum			Adult	Male		Steiner 1968 Excavation	Landesmuseum Sachsen-Anhalt
65	78:1293ff	Caudal vertebra	Upper, incomplete			Adult-Elderly			Steiner 1968 Excavation	Landesmuseum Sachsen-Anhalt
66	78:1293gg	Caudal vertebra	Upper, incomplete			Adult-Elderly			Steiner 1968 Excavation	Landesmuseum Sachsen-Anhalt
67	78:1293hh	Caudal vertebra	Upper, incomplete			Adult-Elderly			Steiner 1968 Excavation	Landesmuseum Sachsen-Anhalt
68	Ma.14309	Caudal vertebra	Upper, incomplete			Adult-Elderly			Excavation Völker 1984/85 (Quadrant F 183)	Museum für Naturkunde Berlin
69	Ma.14310	Caudal vertebra	Middle			Adult-Elderly			Excavation Völker 1984/85 (Quadrant F 187)	Museum für Naturkunde Berlin
70	78:1293ii	Caudal vertebra	Middle			Adult-Elderly			Steiner 1968 Excavation	Landesmuseum Sachsen-Anhalt
71	Ma.14330	Caudal vertebra	Lower			Adult-Elderly			Excavation Steiner Halle/Saale 1962 (81:1225)	Museum für Naturkunde Berlin
72	78:1293jj	Costa	No. 1, incomplete	x		Adult-Elderly			Steiner 1968 Excavation	Landesmuseum Sachsen-Anhalt
73	78:1293kk	Costa	No. 4, upper half		x	Adult-Elderly			Steiner 1968 Excavation	Landesmuseum Sachsen-Anhalt
74	78:1293ll	Costa	No. 3, upper half		x	Adult-Elderly			Steiner 1968 Excavation	Landesmuseum Sachsen-Anhalt
75	78:1293mm	Costa	No. 1-4, distal fragment		x	Adult-Elderly			Steiner 1968 Excavation	Landesmuseum Sachsen-Anhalt

(Table 1) contd.....

No.	Inv.-No.	Bone type	Commentary	Left	Right	Ind. Age	Sex		Old Collection	Collection
76	78:1293nn	Costa	No. 5, upper half		x	Adult-Elderly			Steiner 1968 Excavation	Landesmuseum Sachsen-Anhalt
77	78:1293oo	Costa	No. 3-4, incomplete		x	Adult-Elderly			Steiner 1968 Excavation	Landesmuseum Sachsen-Anhalt
78	78:1293pp	Costa	No. 3.6, distal fragment		x	Adult-Elderly			Steiner 1968 Excavation	Landesmuseum Sachsen-Anhalt
79	Ma.14748	Costa	No. 5	x		Adult-Elderly			Excavation Völker 1984/85 (no Quadrant)	Museum für Naturkunde Berlin
80	Ma. 51522	Costa	No. 6		x	Adult-Elderly			Excavation Völker 1984/85 (Quadrant D86/101)	Museum für Naturkunde Berlin
81	78:1293qq	Costa	Middle rib, distal half			Adult-Elderly			Steiner 1968 Excavation	Landesmuseum Sachsen-Anhalt
82	78:1293rr	Costa	No. 7, distal part missing		x	Adult-Elderly			Steiner 1968 Excavation	Landesmuseum Sachsen-Anhalt
83	78:1293ss	Costa	No. 8, distal part missing			Adult-Elderly			Steiner 1968 Excavation	Landesmuseum Sachsen-Anhalt
84	HH-3	Costa	Middle, distal half	x		Adult-Elderly			Prospections Diedrich, 2012	Rübeland Cave Center
85	78:1293tt	Costa	No. 9, missing distal part, pathology with callus		x	Adult-Elderly			Steiner 1968 Excavation	Landesmuseum Sachsen-Anhalt
86	78:1293uu	Costa	Middle, distal half			Adult-Elderly			Steiner 1968 Excavation	Landesmuseum Sachsen-Anhalt
87	Ma. 51520	Costa	No. 6		x	Adult-Elderly			Excavation Völker 1984/85 (Quadrant F68/69)	Museum für Naturkunde Berlin
88	Ma. 51521	Costa	No. 7		x	Adult-Elderly			Excavation Völker 1984/85 (Quadrant D86)	Museum für Naturkunde Berlin
89	78:1293vv	Costa	No. 11		x	Adult-Elderly			Steiner 1968 Excavation	Landesmuseum Sachsen-Anhalt
90	78:1293ww	Costa	No. 12-13, proximal half		x	Adult-Elderly			Steiner 1968 Excavation	Landesmuseum Sachsen-Anhalt
91	78:1293xx	Costa	No. 12-13, proximal half	x		Adult-Elderly			Steiner 1968 Excavation	Landesmuseum Sachsen-Anhalt
92	78:1293yy	Costa	No. 13		x	Adult-Elderly			Steiner 1968 Excavation	Landesmuseum Sachsen-Anhalt

(Table 1) contd.....

No.	Inv.-No.	Bone type	Commentary	Left	Right	Ind. Age	Sex		Old Collection	Collection
93	Ma.11734	Sternal bone	Middle			Adult-Elderly			Excavation Völker 1984/85 (Quadrant D103)	Museum für Naturkunde Berlin
94	Ma.17736	Sternal bone	Middle			Adult-Elderly			Excavation Völker 1984/85 (Quadrant D85)	Museum für Naturkunde Berlin
95	78:1293xc	Femur	Complete, length = 430 mm, distal width = 90 mm	x		Adult-Elderly	Male		Steiner 1968 Excavation	Landesmuseum Sachsen-Anhalt
96	Ma.14334	Femur	Complete, length = 420 mm, distal width = 87 mm	x		Adult-Elderly	Male		Excavation Völker 1984/85 (Quadrant D 28)	Museum für Naturkunde Berlin
97	Ma.14335	Femur	Without distal joint		x	Adult-Elderly	Male		Excavation Völker 1984/85	Museum für Naturkunde Berlin
98	Ma.9541	Patella	Complete			Adult-Elderly	Male		Excavation Völker 1984/85 (Quadrant D 60)	Museum für Naturkunde Berlin
99	HH-2	Tibia	Distal part			Adult-Elderly	Male		Prospections Diedrich, 2012	Rübeland Cave Center
100	HH-9	Tibia	Without proximal joint			Adult-Elderly	Male		Prospections Diedrich, 2012	Rübeland Cave Center
101	Ma.14333	Tibia	Complete, length = 364 mm, distal width 66 mm	x		Adult-Elderly	Male		Excavation Völker 1984/85 (Quadrant F)	Museum für Naturkunde Berlin
102	78:1293xd	Tibia	Complete, Length = 356 mm, distal width 66 mm		x	Adult-Elderly			Steiner 1968 Excavation	Landesmuseum Sachsen-Anhalt
103	Ma.14324	Fibula	Distal joint	x		Adult-Elderly	Male		Excavation Völker 1984/85	Museum für Naturkunde Berlin
104	78:1293xe	Fibula	Distal end		x	Adult-Elderly	Male		Steiner 1968 Excavation	Landesmuseum Sachsen-Anhalt
105	78:1293xf	Fibula	Distal end		x	Adult-Elderly		Female	Steiner 1968 Excavation	Landesmuseum Sachsen-Anhalt

(Table 1) contd.....

No.	Inv.-No.	Bone type	Commentary	Left	Right	Ind. Age	Sex		Old Collection	Collection
106	78:1293xg	Calcaneus	Complete, length = 124 mm, distal width = 55 mm		x	Adult-Elderly			Steiner 1968 Excavation	Landesmuseum Sachsen-Anhalt
107	78:1293xh	Calcaneus	Complete, length = 134 mm, distal width = 57 mm		x	Adult-Elderly	Male		Steiner 1968 Excavation	Landesmuseum Sachsen-Anhalt
108	78:1293xi	Astragalus	Complete, length = 65 mm, width = 60 mm	x		Adult-Elderly	Male		Steiner 1968 Excavation	Landesmuseum Sachsen-Anhalt
109	78:1293xj	Astragalus	Complete, length = 67 mm, width = 63 mm	x		Adult-Elderly		Female	Steiner 1968 Excavation	Landesmuseum Sachsen-Anhalt
110	78:1293xk	Astragalus	Complete, length = 70 mm, width 66 = mm		x	Adult-Elderly	Male		Steiner 1968 Excavation	Landesmuseum Sachsen-Anhalt
111	78:1293xl	Cuboid	Complete	x		Adult-Elderly	Male		Steiner 1968 Excavation	Landesmuseum Sachsen-Anhalt
112	Ma.14327	Intermedium	Complete	x		Adult-Elderly	Male		Excavation Völker 1984/85 (Quadrant D 107)	Museum für Naturkunde Berlin
113	Ma.14328	Intermedium	Complete		x	Adult-Elderly	Male		Excavation Völker 1984/85 (Quadrant F 184)	Museum für Naturkunde Berlin
114	Ma.14312	Metatarsus	II, complete, length = 142 m, distal width = 26 mm	x		Adult-Elderly		Female	Excavation Völker 1984/85 (Quadrant F 173)	Museum für Naturkunde Berlin
115	78:1293xm	Metatarsus	II, length = 130 mm, distal width = 24 mm			Adult-Elderly	Male		Steiner 1968 Excavation	Landesmuseum Sachsen-Anhalt
116	78:1293xn	Metatarsus	II, length = 139 mm, distal width = 27 mm			Adult-Elderly	Male		Steiner 1968 Excavation	Landesmuseum Sachsen-Anhalt

(Table 1) contd.....

No.	Inv.-No.	Bone type	Commentary	Left	Right	Ind. Age	Sex		Old Collection	Collection
117	78:1293xo	Metatarsus	II, length = 138 mm, distal width = 29 mm			Adult-Elderly	Male		Steiner 1968 Excavation	Landesmuseum Sachsen-Anhalt
118	Ma.14313	Metatarsus	III, complete, length = 132 mm, distal width = 25 mm	x		Adult-Elderly		Female	Excavation Völker 1984/85 (Quadrant D 176)	Museum für Naturkunde Berlin
119	Ma.14315	Metatarsus	III, complete, length 136 m, distal width 26 mm	x		Adult-Elderly	Male		Excavation Völker 1984/85 (Quadrant F 170)	Museum für Naturkunde Berlin
120	78:1293xp	Metatarsus	IV, complete, length = 154 mm, distal width = 27 mm	x		Adult-Elderly	Male		Steiner 1968 Excavation	Landesmuseum Sachsen-Anhalt
121	78:1293xq	Metatarsus	IV, complete, length = 151 mm, distal width = 27 mm		x	Adult-Elderly	Male		Steiner 1968 Excavation	Landesmuseum Sachsen-Anhalt
122	Ma.14316	Metatarsus	IV, complete, length = 143 m, distal width = 24 mm		x	Adult-Elderly		Female	Excavation Völker 1984/85 (Quadrant D 140)	Museum für Naturkunde Berlin
123	Ma.14314	Metatarsus	V, complete, length = 137 m, distal width = 22 mm	x		Adult-Elderly		Female	Excavation Völker 1984/85 (Quadrant D 105)	Museum für Naturkunde Berlin
124	Ma.14712	Phalanx I	Incomplete			Adult-Elderly			Excavation Völker 1984/85	Museum für Naturkunde Berlin
125	Ma.9869	Phalanx I	Complete, ?digit V		x	Adult-Elderly			Excavation Völker 1984/85 (Quadrant F 171)	Museum für Naturkunde Berlin
126	78:1293xa	Phalanx I	Complete, ?digit II		x	Adult-Elderly			Steiner 1968 Excavation	Landesmuseum Sachsen-Anhalt
127	78:1293xr	Phalanx II	Complete, ?digit IV		x	Adult-Elderly			Steiner 1968 Excavation	Landesmuseum Sachsen-Anhalt
128	78:1293xb	Phalanx III	Complete			Adult-Elderly			Steiner 1968 Excavation	Landesmuseum Sachsen-Anhalt

Table 2. Metrics of compared European Pleistocene lion crania (cf. Fig. 12, data from [4, 14, 15, 27 - 34, 36] with updated and new data from the Upper Rhine Graben and improved dating).

Locality	Dating	Length (mm)	Condylus Width (mm)
Edingen (Brühl) (D) - skeleton	Weichselian/Wuermian Glacial, Upper terrace, MIS 3	440	75
Bottrop (D)	Weichselian/Wuermian Glacial, MIS 3-5d	420	74
Azé Cave (Fr)	Late Saalian/Rissian, MIS 6-8	417	74
Huttenheim (D) - skeleton	Weichselian/Wuermian Glacial, Lower terrace, MIS 3-5d	410	74
Arrikrutz skeleton (Es)	Late Saalian/Rissian, MIS 6-8	407	72
Zoolithen Cave (D) - holotype	Weichselian/Wuermian Glacial, MIS 3-5d	402	72
Bobenheim-Roxheim (D)	Weichselian/Wuermian Glacial, MIS 3-5d	390	73
Vypustek Cave - skeleton Vienna (CZ)	Weichselian/Wuermian Glacial, MIS 3-5d	390	73
Siegsdorf open air skeleton (D)	Weichselian/Wuermian Glacial, MIS 3-5d	384	71
Sloup Cave (CZ)	Weichselian/Wuermian Glacial, MIS 3-5d	378	72
Vypustek Cave - skeleton Brno (CZ)	Weichselian/Wuermian Glacial, MIS 3-5d	357	68
Zandibbio (It)	Eemian Interglacial, MIS 5e	324	60
Neumark-Nord Lake 1 (D) - skeleton	Eemian Interglacial, MIS 5e	(313) reconstructed	65
Srbsko Chlum Komin Cave (CZ) - skeleton (subadult)	Weichselian/Wuermian Glacial, MIS 3-5d	305	65
Hýskov (CZ)	Cromerian Complex, MIS 11-21	302	65
Perick Caves (D)	Eemian Interglacial, MIS 5e	301	66
Urşilor Cave skeleton (Ro)	Eemian Interglacial, MIS 5e	282	65
Scocu Scorotei Cave (Ro)	Eemian Interglacial, MIS 5e	275	67

Table 3. Metrics of compared European Pleistocene lion humeri (cf. Fig. 12 data from [4, 14, 15, 27 - 34, 36] with updated and new data from the Upper Rhine Graben and improved dating).

Locality	Dating	Length (mm)	Distal Width (mm)
Hermann's Cave (D)	Weichselian/Wuermian Glacial, MIS 3-5d	396	104
Urşilor Cave, skeleton 1 (RO)	Weichselian/Wuermian Glacial, MIS 3-5d	396	100
Altlußheim-Eichelgarten (D)	Weichselian/Wuermian Glacial, MIS 3-5d	386	102
Hermann's Cave (D)	Weichselian/Wuermian Glacial, MIS 3-5d	386	100
Huttenheim (D) skeleton	Weichselian/Wuermian Glacial, MIS 3-5d	385	96
Hermann's Cave (D)	Weichselian/Wuermian Glacial, MIS 3-5d	381	100
Vypustek Cave,1 (CZ)	Weichselian/Wuermian Glacial, MIS 3-5d	381	100
Vypustek Cave, 2 (CZ)	Weichselian/Wuermian Glacial, MIS 3-5d	379	100
Hermann's Cave (D)	Weichselian/Wuermian Glacial, MIS 3-5d	376	94
Bruehl, skeleton (D)	Weichselian/Wuermian Glacial, MIS 3-5d	375	96
Arrikrutz Cave, skeleton (Es)	Late Saalian/Rissian, MIS 6-8	371	103
Brühl-Rheingewann (D)	Weichselian/Wuermian Glacial, MIS 3-5d	371	97
Zoolithen Cave, 1 (D)	Weichselian/Wuermian Glacial, MIS 3-5d	364	92
Vypustek Cave, 3 (CZ)	Weichselian/Wuermian Glacial, MIS 3-5d	362	95
Keppler Cave, 3 (D)	Weichselian/Wuermian Glacial, MIS 3-5d	362	95
Keppler Cave, 2 (D)	Weichselian/Wuermian Glacial, MIS 3-5d	360	98
Zoolithen Cave, 2 (D)	Weichselian/Wuermian Glacial, MIS 3-5d	360	94
Crumstadt (D)	Weichselian/Wuermian Glacial, MIS 3-5d	358	82
Otterstadt-Waldwiesen (D)	Eemian Interglacial,MIS 5e	357	89
Neumark-Nord Lake 1 (D)	Eemian Interglacial, MIS 5e	362	92
Baumann's Cave (D)	Eemian or Holsteinian Interglacial, MIS 5e/9	343	87
Keppler Cave, 1 (D)	Eemian Interglacial, MIS 5e	342	85
Zoolithen Cave, 3 (D)	Eemian or Holsteinian Interglacial, MIS 5e/9	338	80
Huttenheim-Sandfeld (D)	Eemian Interglacial, MIS 5e	330	84
Bensheim (D)	Eemian Interglacial, MIS 5e	321	82
Srbsko Chlum Komin Cave (CZ) - skeleton (subadult)	Weichselian/Wuermian Glacial, MIS 3-5d	319	84

Table 4. Metrics of compared European Pleistocene lion ulnae (cf. Fig. 12 data from [4, 14, 15, 27 - 34, 36] with updated and new data from the Upper Rhine Graben and improved dating).

Locality	Dating	Length (mm)	Distal Width (mm)
Crumstadt (D)	Weichselian/Wuermian Glacial, MIS 3-5d	440	78
Edingen (D) - skeleton	Weichselian/Wuermian Glacial, MIS 3	432	84
Ketsch-Hohwiesen (D)	Weichselian/Wuermian Glacial, MIS 3-5d	423	72
Zoolithen Cave (D)	Weichselian/Wuermian Glacial, MIS 3-5d	421	77
Großrohrheim (D)	Weichselian/Wuermian Glacial, MIS 3-5d	417	69
Perick Caves, 1 (D)	Weichselian/Wuermian Glacial, MIS 3-5d	410	82
Hermann's Cave (D)	Weichselian/Wuermian Glacial, MIS 3-5d	409	68
Bilstein Cave (D)	Weichselian/Wuermian Glacial, MIS 3-5d	406	71
Huttenheim (D) - skeleton	Weichselian/Wuermian Glacial, MIS 3-5d	404	69
Perick Caves, 2 (D)	Weichselian/Wuermian Glacial, MIS 3-5d	369	79
Neumark-Nord Lake 1 (D) - skeleton	Eemian Interglacial, MIS 5e	364	67
Bottrop (D)	Eemian Interglacial, MIS 5e	357	61
Srbsko Chlum Komin Cave (CZ) - skeleton (subadult)	Weichselian/Wuermian Glacial, MIS 3-5d	348	60
Niederlehme (D)	Eemian Interglacial, MIS 5e	347	60

Table 5. Metrics of compared Central European Pleistocene lion radii (cf. Fig. 12 data from [4, 14, 15, 27 - 34, 36] with updated and new data from the Upper Rhine Graben and improved dating).

Locality	Dating	Length (mm)	Distal Width (mm)
Zoolithen Cave (D)	Weichselian/Wuermian Glacial, MIS 3-5d	36.40	7.40
Wattenheim (D)	Weichselian/Wuermian Glacial, MIS 3-5d	35.80	7.50
Otterstadt-Waldwiesen (D)	Weichselian/Wuermian Glacial, MIS 3-5d	35.80	7.60
Bruehl (D) - skeleton	Weichselian/Wuermian Glacial, MIS 3-5d	35.10	7.60
Geddin (D)	Weichselian/Wuermian Glacial, MIS 3-5d	32.80	7.30
Ursilor Cave (ro) skeleton	Weichselian/Wuermian Glacial, MIS 3-5d	34.20	6.60
Výpustek Cave, 1 (CZ)	Weichselian/Wuermian Glacial, MIS 3-5d	34.00	7.50
Hermann's Cave (D)	Weichselian/Wuermian Glacial, MIS 3-5d	33.90	7.00
Hermann's Cave (D)	Weichselian/Wuermian Glacial, MIS 3-5d	33.80	7.00

(Table 5) contd.....

Locality	Dating	Length (mm)	Distal Width (mm)
Bensheim (D)	Eemian Interglacial, MIS 5e	33.70	6.50
Huttenheim (D) - skeleton	Weichselian/Wuermian Glacial, MIS 3-5d	33.20	7.00
Výpustek Cave, 2 (CZ)	Eemian Interglacial, MIS 5e	32.40	6.80
Bottrop (D)	Eemian Interglacial, MIS 5e	31.70	6.40
Neumark-Nord Lake 1 (D)	Eemian Interglacial, MIS 5e	30.70	7.30
Trmice (CZ)	Eemian Interglacial, MIS 5e	30.50	6.60
Neumark-Nord Lake 1 (D) - skeleton	Eemian Interglacial, MIS 5e	29.40	6.90
Srbsko Chlum Komin Cave (CZ) - skeleton (subadult)	Eemian Interglacial, MIS 5e	28.80	6.40
Niederlehme (D)	Eemian Interglacial, MIS 5e	27.80	5.60

Table 6. Metrics of compared European Pleistocene lion femora (cf. Fig. 12 data from [4, 14, 15, 27 - 34, 36] with updated and new data from the Upper Rhine Graben and improved dating).

Locality	Dating	Length (mm)	Distal Width (mm)
Crumstadt (D)	Weichselian/Wuermian Glacial, MIS 3-5d	497	95
Edingen (D) - skeleton	Weichselian/Wuermian Glacial, MIS 3	460	96
Hermann's Cave (D)	Weichselian/Wuermian Glacial, MIS 3-5d	430	90
Vypustek Cave, 1 (CZ)	Weichselian/Wuermian Glacial, MIS 3-5d	420	91
Vypustek Cave, 2 (CZ)	Weichselian/Wuermian Glacial, MIS 3-5d	425	89
Zoolithen Cave, 1 (D)	Weichselian/Wuermian Glacial, MIS 3-5d	422	79
Zoolithen Cave, 2 (D)	Weichselian/Wuermian Glacial, MIS 3-5d	421	80
Hermann's Cave (D)	Weichselian/Wuermian Glacial, MIS 3-5d	420	87
Geinsheim (D)	Weichselian/Wuermian Glacial, MIS 3-5d	418	78
Crumstadt (D)	Weichselian/Wuermian Glacial, MIS 3-5d	412	80
Siegsdorf (D) - skeleton	Weichselian/Wuermian Glacial, MIS 3-5d	410	86
Keppler Cave (D)	Weichselian/Wuermian Glacial, MIS 3-5d	405	86
Neumark-Nord Lake 1 (D) - skeleton	Eemian Interglacial, MIS 5e	403	86

(Table 6) contd.....

Locality	Dating	Length (mm)	Distal Width (mm)
Zoolithen Cave, 3 (D)	Eemian or Holsteinian Interglacial, MIS 5e/9	401	77
Großrohrheim (D)	Eemian Interglacial, MIS 5e	394	84
Heddesheim (D)	Eemian Interglacial, MIS 5e	385	83
Heddesheim-Neuwiesen (D)	Eemian Interglacial, MIS 5e	385	83
Výpustek Cave, 3 (CZ)	Eemian Interglacial, MIS 5e	380	83
Srbsko Chlum Komin Cave (CZ) - skeleton (subadult)	Eemian Interglacial, MIS 5e	359	84

Table 7. Metrics of compared European Pleistocene lion tibiae (cf. Fig. 12 data from [4, 14, 15, 27 - 34, 36] with updated and new data from the Upper Rhine Graben and improved dating).

Locality	Dating	Length (mm)	Distal Width (mm)
Edingen (D) - skeleton	Weichselian/Wuermian Glacial, MIS 3	393	74
Balve Cave (D)	Weichselian/Wuermian Glacial, MIS 3-5d	392	67
Arrikrutz Cave (Es) - skeleton	Late Saalian/Rissian, MIS 6-8	370	69
Huttenheim (D) - skeleton	Weichselian/Wuermian Glacial, MIS 3-5d	367	67
Praha-Podbaba (D)	Weichselian/Wuermian Glacial, MIS 3-5d	364	67
Hermann's Cave (D)	Weichselian/Wuermian Glacial, MIS 3-5d	364	66
Výpustek Cave, 1 (CZ)	Weichselian/Wuermian Glacial, MIS 3-5d	364	68
Výpustek Cave, 2 (CZ)	Weichselian/Wuermian Glacial, MIS 3-5d	362	69
Bruehl (D) - skeleton	Weichselian/Wuermian Glacial, MIS 3-5d	358	69
Bielstein Cave (D)	Weichselian/Wuermian Glacial, MIS 3-5d	358	61
Hermann's Cave (D)	Weichselian/Wuermian Glacial, MIS 3-5d	356	66
Siegsdorf (D) - skeleton	Weichselian/Wuermian Glacial, MIS 3-5d	355	71
Keppler Cave (D)	Eemian Interglacial, MIS 5e	355	60
Zoolithen Cave, 3 (D)	Eemian Interglacial, MIS 5e	354	60
Zoolithen Cave, 4 (D)	Eemian Interglacial, MIS 5e	353	56

(Table 7) contd.....

Locality	Dating	Length (mm)	Distal Width (mm)
Bensheim (D)	Eemian Interglacial, MIS 5e	346	62
Výpustek Cave, 3 (CZ)	Eemian Interglacial, MIS 5e	345	61
Zoolithen Cave, 5 (D)	Eemian or Holsteinian Interglacial, MIS 5e/9	344	57
Speyer-Binstetal-Südost (D)	Eemian Interglacial, MIS 5e	340	61
Zoolithen Cave, 6 (D)	Eemian or Holsteinian Interglacial, MIS 5e/9	338	57
Neumark-Nord Lake 1 (D) - skeleton	Eemian Interglacial, MIS 5e	336	65
Bensheim (D)	Eemian Interglacial, MIS 5e	332	63
Kleinenbesten (D)	Eemian Interglacial, MIS 5e	322	61

REFERENCES

[1] Goldfuss GA. Die Umgebungen von Muggendorf. Erlangen: Ein Taschenbuch für Freunde der Natur und Altertumskunde 1810; p. 351.

[2] Goldfuss GA. Osteologische Beiträge zur Kenntnis verschiedener Saeugethiere der Vorwelt: IV. Ueber den Schaedel des Hoehlenloewen. Verh Kais Leopold Carol Akad Natur 1821; 10(2): 489-94.

[3] Diedrich C. The rediscovered holotypes of the Upper Pleistocene spotted hyena *Crocuta crocuta spelaea* (Goldfuss, 1823) and the steppe lion *Panthera leo spelaea* (Goldfuss, 1810) and taphonomic discussion to the Geilenreuther Cave hyena den (South-Germany). Zool J Linn Soc London 2008; 154: 822-31.
[http://dx.doi.org/10.1111/j.1096-3642.2008.00425.x]

[4] Diedrich C. The largest European lion *Panthera leo spelaea* (Goldfuss) population from the Zoolithen Cave, Germany – specialized cave bear predators of Europe. Hist Biol 2011; 23(2/3): 271-311.
[http://dx.doi.org/10.1080/08912963.2010.546529]

[5] Diedrich C. Holotype skulls, stratigraphy, bone taphonomy and excavation history in the Zoolithen Cave and new theory about Esper's "great deluge". Quat Sci J 2014; 63(1): 78-98.

[6] Burger J, Rosendahl W, Loreille O, *et al.* Molecular phylogeny of the extinct cave lion *Panthera leo spelaea.* Mol Phylogenet Evol 2004; 30(3): 841-9.
[http://dx.doi.org/10.1016/j.ympev.2003.07.020] [PMID: 15012963]

[7] Barnett R, Shapiro B, Barnes I, *et al.* Phylogeography of lions (Panthera leo ssp.) reveals three distinct taxa and a late Pleistocene reduction in genetic diversity. Mol Ecol 2009; 18(8): 1668-77.
[http://dx.doi.org/10.1111/j.1365-294X.2009.04134.x] [PMID: 19302360]

[8] Diedrich C. Palaeopopulations of Late Pleistocene top predators in Europe: Ice Age spotted hyenas and steppe lions in battle and competition about prey. Paleont J 2014; pp. 1-34.

[9] Bächler E. Das Wildkirchli. Eine Monographie. Buchdruckerei H. Tschudy 1936; p. 254.

[10] Diedrich C. Late Pleistocene leopards across Europe - most northern European population, highest elevated records in the Alps, complete skeletons in the Dinarids and comparison to the Ice Age cave art. Quat Sci Rev 2013; 76: 167-93. [http://dx.doi.org/10.1016/j.quascirev.2013.05.009]

[11] Dawkins WD, Sandfort A. A monograph of the British Pleistocene Mammalia. Volume I. The British Pleistocene Felidae. Palaeont Soc Monogr 1900; 1866-1872: 1-193.

[12] Altuna J. Fund eines Skelettes des Höhlenlöwen (*Panthera leo spelaea* Goldfuss) in Arrikrutz, Baskenland. Bonner zool Beitr 1981; 32(32(1/2)): 31-46.

[13] Tichy G. Über der Fund eines Höhlenlöwen (*Panthera felis spelaea* Goldfuss) aus dem Tennengebirge bei Salzburg. Mitt Ges Salzburgische Landesk 1985; 125: 845-64.

[14] Argant A. Étude de 'exemplaire de *Panthera spelaea* (Goldfuss, 1810) (Mammalia, Carnivora, Felidae) du gisement Pleistocène moyen recent de la grotte d" Aze (Saone et Loire). Rev Paleobiol 1988; 7(2): 449-66.

[15] Gross C. Das Skelett des Höhlenlöwen (*Panthera leo spelaea* Goldfuss, 1810) aus Siegsdorf/Ldkr. Traunstein im Vergleich mit anderen Funden aus Deutschland und den Niederlanden. Unpublished Dissertation, Tierärztliche Fakultät der Maximilians-Universität, Munich 1992; 129.

[16] Fischer K. Neufunde von jungpleistozänen Höhlenlöwen *Panthera leo spelaea* (Goldfuss, 1810) in Rübeland (Harz). Braunschweiger naturk Schr 1994; 4(3): 455-71.1994;

[17] Guzvica G. *Panthera spelaea* (Goldfuss, 1810) from North-Western Croatia. Geol Croat 1998; 51(1): 7-14.

[18] Baryshnikov GF, Boeskorov G. The Pleistocene cave lion, *Panthera spelaea* (Carnivora, Felidae) from Yakutia, Russia. Cranium 2001; 18: 7-4.

[19] Bona F. Systematic position of a complete lion-like cat skull from the Eemian ossiferous rubble near Zandobbio (Bergamo, North Italy). Riv Ital Paleontol Stratigr 2006; 112(1): 157-66.

[20] Barycka E. Middle and Late Pleistocene Felidae and Hyaenidae of Poland Fauna Poloniae Fauna Polski 2ns. Warszawa: Mus Inst Zool Polish Acad Sci 2008; p. 128.

[21] Marciszak A, Stefaniak K. Two forms of cave lion: Middle Pleistocene *Panthera spelaea fossilis* Reichenau, 1906 and Upper Pleistocene *Panthera spelaea spelaea* (Goldfuss, 1810) from the Bísník Cave, Poland. N Jb Geol Paläont Abh 2010; 258(3): 339-51.

[22] Sabol M. A record of Pleistocene lion-like felids in the territory of Slovakia. Quat Hors 2011; 4: 215-28.

[23] Ballesio R. Étude de *Panthera* (*leo*) *spelaea* (Goldfuss) nov. subsp. (Mammalia, Carnivora, Felidae) du gisement du Pleistocène moyen des Abimes de la Fage à Noailles (Corrèze). Nouv Archeol Mus Hist Nat Lyon 1975; 13: 47-55.

[24] Boule M. Les grands chats des caverns. Ann Paleontol 1906; 1: 69-95.

[25] Dietrich WO. Fossile Löwen im europäischen und afrikanischen Pleistozän. Paläont Abh A Paläozool 1968; 3(2): 323-66.

[26] Rathgeber T. Neue jungpleistozäne Säugetier-Reste aus der Aufhauser Höhle (Kat.-Nr. 7424/13) bei Geislingen an der Steige (Schwäbische Alb). Mitt Verb dt Höhlen- Karstf 1982; 28(1): 9-12.

[27] Diedrich C. Upper Pleistocene *Panthera leo spelaea* (Goldfuss, 1810) skeleton remains from Praha-Podbaba and contribution to other lion finds from loess and river terrace sites in Central Bohemia (Czech Republic). Bull Geosci 2007; 82(2): 99-117. [http://dx.doi.org/10.3140/bull.geosci.2007.02.99]

[28] Diedrich C. Upper Pleistocene *Panthera leo spelaea* (Goldfuss, 1810) remains from the Bilstein Caves (Sauerland Karst) and contribution to the steppe lion taphonomy, palaeobiology and sexual dimorphism. Ann Paleontol 2009; 95: 117-38.

[http://dx.doi.org/10.1016/j.annpal.2009.07.001]

[29] Diedrich C. Steppe lion remains imported by Ice Age spotted hyenas into the Late Pleistocene Perick Caves hyena den in Northern Germany. Quat Res 2009; 71(3): 361-74. [http://dx.doi.org/10.1016/j.yqres.2008.12.006]

[30] Diedrich C. A diseased *Panthera leo spelaea* (Goldfuss, 1810) lioness from a forest elephant graveyard in the Late Pleistocene (Eemian) interglacial lake at Neumark-Nord, central Germany. Hist Biol 2010; 2010(4): 1-23.

[31] Diedrich C. A Late Pleistocene lion *Panthera leo spelaea* (Goldfuss, 1810) skull and other postcranial remains from the Sloup Cave in the Moravian Karst, Czech Republic. Acta Archaeol 2011; 19-28.

[32] Diedrich C. Late Pleistocene steppe lion *Panthera leo spelaea* (Goldfuss, 1810) footprints and bone remains from open air sites in northern Germany – evidence of hyena-lion antagonism in Europe. Quat Sci Rev 2011; 30: 1883-906. [http://dx.doi.org/10.1016/j.quascirev.2011.03.006]

[33] Diedrich C. Pleistocene *Panthera leo spelaea* (Goldfuss, 1810) remains from the Balve Cave (NW Germany) – a cave bear, hyena den and Middle Palaeolithic human cave, and review of the Sauerland Karst lion sites. Quaternaire 2011; 22(2): 105-27. [http://dx.doi.org/10.4000/quaternaire.5897]

[34] Diedrich C. The Late Pleistocene *Panthera leo spelaea* (Goldfuss, 1810) skeletons from the Sloup and Srbsko Caves in Czech Republic (central Europe) and contribution to steppe lion cranial pathologies and postmortally damages as results of interspecies fights, hyena antagonism and cave bear attacks. Bull Geosci 2011; 86(4): 817-40. [http://dx.doi.org/10.3140/bull.geosci.1263]

[35] Diedrich C. Ice Age geomorphological Ahorn Valley and Ailsbach River terrace evolution– and its importance for the cave use possibilities by cave bears, top predators (hyenas, wolves and lions) and humans (Late Magdalénians) in the Frankonia Karst – case studies in the Sophie's Cave near Kirchahorn, Bavaria. Quat Sci J 2013; 62(2): 162-74.

[36] Diedrich C, Rathgeber T. Late Pleistocene steppe lion *Panthera leo spelaea* (Goldfuss, 1810) skeleton remains of the Upper Rhine valley (SW Germany) and contribution to their sexual dimorphism, taphonomy and habitus. Hist Biol 2012; 24(1): 1-28. [http://dx.doi.org/10.1080/08912963.2010.549943]

[37] Reichenau W. Beiträge zur näheren Kenntnis der Carnivoren aus den Sanden von Mauer und Mosbach. Abhandlungen der Großherzoglich Hessischen Geologischen Landesanstalt zu Darmstadt 1906; 4(2): 189-313.

[38] Schütt G, Hemmer H. Zur Evolution des Lowen (*Panthera leo* L.) im europaischen Pleistozan. N Jb Geol Palaont Mh 1978; 4: 228-4.

[39] Schütt G. Untersuchungen am Gebiß von Panthera leo fossilis (Reichenau, 1906) und Panthera leo spelaea (Goldfuss, 1810). N Jb Geol Paläont Abh 1969; 134: 192-220.

[40] Diedrich C. Mid/Late Pleistocene large glacial and small interglacial lion subspecies from the Rhine Valley Ice Age Safari migratory channel. In: Ice Age Safari along the Rhine Valley migratory channnel, Chapter IV Predators. Elsevier (accepted for 2017).

[41] Diedrich C, McFarlane DA. The Saber tooth cat (*Homotherium*) from Middle Pleistocene archaeological and carnivore den sites of Germany - taxonomy, taphonomy and a revision of the Schöningen felids. Quat Int 2017; 436: 76-83. [http://dx.doi.org/10.1016/j.quaint.2016.10.015]

[42] Schaller G. The Serengeti Lion A Study of Predator-Prey Relations. Chicago: University of Chicago Press 1972; p. 494.

[43] Joubert D. Hunting behaviour of lions (*Panthera leo*) on elephants (*Loxodonta africana*) in the Chobe National Park, Botswana. Afr J Ecol 200; 644(2): 279-81.

[44] Hayward W, Kerley GI. Prey preferences of the lion (*Panthera leo*). J Zool (Lond) 2005; 267: 309-22. [http://dx.doi.org/10.1017/S0952836905007508]

[45] Estes R. The safari companion: a guide to watching African mammals. Vermont: Chelsea Green Publishing Company 1999; p. 459.

[46] Diedrich C. Late Pleistocene European Ice Age spotted hyena palaeopathologies from inter-/intraspecies battles with top predators. Int J Paleopathol 2017.

[47] Diedrich C. Steppe lion predation – another reason for cave bears to protect during hibernation against carnivores in deepest parts of large cave systems in Europe. Abstract 15th International Cave Bear Symposium, Spisska Nova Ves, Slovakia, 2009.

[48] Diedrich C. Cave bear killers and scavengers from the last Ice Age of central Europe: Feeding specializations in response to the absence of mammoth steppe fauna from mountainous regions. Quat Int 2012; 255: 59-78. [http://dx.doi.org/10.1016/j.quaint.2011.06.048]

[49] Rothschild BM, Diedrich C. Comparison of pathologies in the extinct Pleistocene Eurasian steppe lion *Panthera leo spelaea* (Goldfuss, 1810) to those in the modern lion, *Panthera leo* - Results of fights with hyenas, bears and lions and other ecological stress. Int J Paleopathol 2012; 2(4): 187-98. [http://dx.doi.org/10.1016/j.ijpp.2012.09.004]

[50] Kruuk H. The spotted hyena A story of predation and social behavior. Chicago: The University of Chicago Press 1972.

[51] Münzel SC, Stiller M, Hofreiter M, Mittnik A, Conard NJ, Bocherens H. Pleistocene bears in the Swabian Jura (Germany): Genetic replacement, ecological displacement, extinctions and survival. Quat Int 2011; 245: 225-37. [http://dx.doi.org/10.1016/j.quaint.2011.03.060]

[52] Diedrich C. Ichnological and ethological studies in one of Europe's famous bear den in the Urşilor Cave (Carpathians, Romania). Ichnos 2011; 18(1): 9-26. [http://dx.doi.org/10.1080/10420940.2011.552578]

[53] Bocherens H, Drucker DG, Bonjean D, *et al.* Isotopic evidence for dietary ecology of cave lion (*Panthera spelaea*) in North-Western Europe: Prey choice, competition and implications for extinction. Quat Int 2011; 245: 249-61. [http://dx.doi.org/10.1016/j.quaint.2011.02.023]

[54] Diedrich C. Sophie's Cave – a Late Pleistocene cave bear den. Bentham/Elsevier 2015; p. 192.

[55] Argenti P, Mazza PP. Mortality analyses of the Late Pleistocene bears from Grotta Lattaia, central Italy. J Arch Sci 2006; 33: 1552-8. [http://dx.doi.org/10.1016/j.jas.2006.02.007]

[56] Debeljak D. Fossil population structure and mortality of the cave bear from the Mikrica Cave (North Slovenia). Acta Carsol 2007; 36(3): 475-84. [http://dx.doi.org/10.3986/ac.v36i3.181]

[57] Grandal-D' Anglade A, Vidal Romani JR. A population study of the cave bear (*Ursus spelaeus* Ros.-Hein.) from Cova Eiros (Triacastela, Galicia, Spain). Geobios 1997; 30(5): 723-31. [http://dx.doi.org/10.1016/S0016-6995(97)80160-X]

[58] Rabeder G, Nagel D, Pacher M. Der Höhlenbär Thorbecke Species 4. Stuttgart: Thorbecke 2000; p. 111.

[59] Kurtèn B. The cave bear story. Columbia University Press, New York, New York 2076; p. 163.

CHAPTER 5

Ice Age Wolf Den and Spotted Hyena Impact

Abstract: Few Late Pleistocene Ice Age spotted hyena *Crocuta crocuta spelaea* (Goldfuss, 1823) bones (18% of NISP) of possibly a single subadult individual from the Hermann's Cave, Harz Mountain Range, Northern Germany indicates a very sporadic short term use possibly as a natal/birth den. Few (2% of NISP) mammoth *Mammuthus primigenius* and woolly rhinoceros *Coelodonta antiquitatis* (7% of NISP) guilt seem to have been imported and chew damaged by hyenas. The large amount (43% of NISP) of Ice Age wolf *Canis lupus spelaeus* (Goldfuss, 1823) remains suggest the identification of a wolf den (natal den) in the anterior cave area, in which individual population had at least one elderly male, one female and one subadult (based on the femorae and cranial remains). The guilt of wolf activities is the holarctic reindeer *Rangifer tarandus* (23% of NISP), but also alpine chamios *Rupicapra rupicapra* (5% of NISP). The same former entrance area was used possibly as very short-time hyen natal den. Other Late Pleistocene Ice Age spotted hyena and wolf dens are located within a 50 km radius mainly north of the Harz Mountain Range. All those dens are situated along rivers (several in gypsum karst doline areas or on river terraces). Those valleys/terraces must have been used by reindeers and other megafauna for seasonal summer migrations along the streams into the Harz Mountains. As known for several other European caves, both large predators and scavengers are responsible for most of the cave bear carcass dismembering and bone damages. This is found especially true for most of the cave bear cub bones, whereas only hyenas left typical non-cracked and by the premolar crushing teeth oval-hole punctured cave bear cub femorae (= pseudo-bone flutes), which seem to have dwelt regularly the Hermann's Cave for cave bear carcass feeding.

Keywords: Late Pleistocene, Hyena *Crocuta crocuta spelaea*, Wolf *Canis lupus spelaeus* den, Guilt, Specialized cave bear feeding in mountain regions of Europe.

ICE AGE SPOTTED HYENA AND WOLF RECORD IN THE HARZ MOUNTAIN REGION

The Late Pleistocene lion antagonist predators was the Ice Age spotted hyena *Crocuta crocuta spelaea* (Goldfuss, 1823) [1], which sites, especially their dens, have been mapped recently in Germany [2]. Most den sites are caves, whereas more and more open air bone accumulation sites can be referred to hyena activity or even communal den sites [2 - 4]. The Late Pleistocene Ice Age wolves *Canis lupus spelaeus* (Goldfuss, 1823) [1] are still not reviewed, nor is a new study

Cajus G. Diedrich

available for German sites. Such a revision of a "specialized ecomorph wolf" of Europe [5] is of need. Also this species is based on a skull from the German Zoolithen Cave [6]. The only known footprints from the Late Pleistocene (MIS 3) have been mapped in-between several other megafauna trackways (mostly reindeer) at the NW-German open air Emscher River terrace site Bottrop [3]. In the Harz Mountains, the few frequented/short-term used Late Pleistocene spotted hyena dens were the Unicorn Cave and the Hermann's Cave [7] (Fig. **1A**). The hyena bone-rich predator den sites are surrounding the Harz slopes in the North with the Seweckenberge and the Westeregeln gypsum karst sites [8, 9] (Fig. **1A**). The Königsaue river terrace site has delivered only a few hyena and guilt remains [9]. On the southern Harz Mountain slopes there are hyena dens especially in the doline-rich gypsum karst belt at Osterode or Sangerhausen [9], whereas the herein new added den site Bennungen lies along the Helme stream terrace (Fig. **1A**).

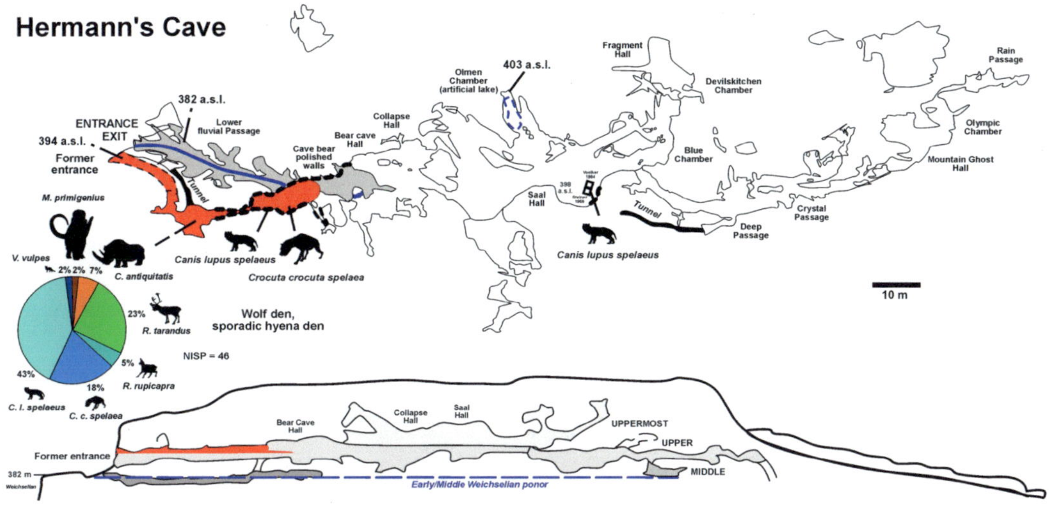

Figure 1. Hermann's Cave wolf and sporadic short-term spotted hyena den within the first 20 meters of the former entrance (modified from [7], and updated in the fauna NISP).

Ice Age Wolf Remains from the Hermann's Cave

17 bones from *Canis lupus spelaeus* (Goldfuss, 1823) (former "cave wolf") [1] (Table **1**) are from the anterior part of the Hermann's Cave (Fig. **1B**). Those were found in the first meters of the Bear Cave Hall and two carnivore cracked femorae are from the Saal Hall [7]. First descriptions of mixed wolf bones with other canid material from the opposite Bauman's Cave [10], have recently been separated and revised excluding also the Middle Pleistocene wild dog bones [7].

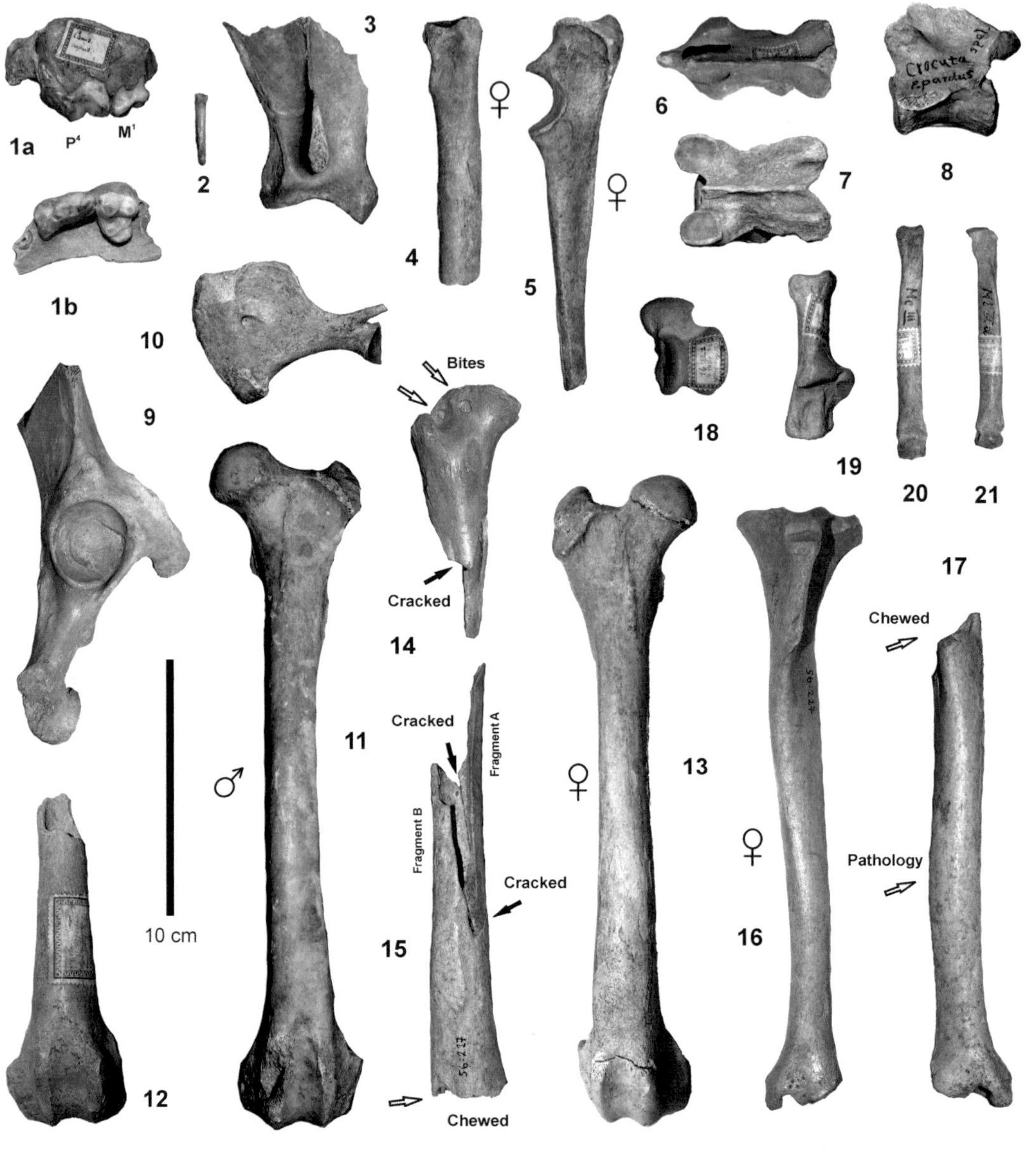

Figure 2. Late Pleistocene Ice Age wolf *Canis lupusspelaeus* (Goldfuss, 1823) remains from the Hermann's Cave, Rübeland, Saxony-Anhalt (Germany). 1. Left maxillary from the Bear Hall (LMNB), a. Lateral, b. Occlusal. 2. Left I_2 from the Bear Hall (LMNB), lingual. 3. Right scapula from the Bear Hall (RH), lateral. 4. Right ulna from the Bear Hall (LMNB), cranial. 5. Right radius (LMNB), lateral. 6. Axis from the Bear Hall (LMNB), dorsal. 7. Cervical vertebra (C3) from the Bear Hall (LMNB), dorsal. 8. Lumbal vertebra from the Bear Hall (LMNB), lateral. 9. Coxa from the Bear Hall (LMNB), lateral. 10. Ischium from the Lower Fluvial Passage cone (RH), lateral. 11. Left Femur from the Collapse Hall (RH), cranial. 12. Left distal femur from the Bear Hall (LMNB), cranial. 13. Left femur from the Bear Hall (LMNB), cranial. 14. Carnivore damaged left femur from the Bear Hall (LDA), caudal. 15. Carnivore crack-damaged right femur from the Saal Hall (LDA), cranial. 16. Right tibia from the Saal Hall (LDA), cranial. 17. Left distal tibia proximally chewed

from the Bear Hall (RH), cranial. 18. Right astragal from the Bear Hall (LMNB), dorsal. 19. Right calcaneus the Bear Hall (LMNB), cranial. 20. Right Mc III the Bear Hall (LMNB), dorsal. 21. Right Mt IV the Bear Hall (LMNB), dorsal. Composed from [7] and modified.

The few wolf cranial material was compared to published skulls from the Zoolithen Cave [6], or English cave sites [11]. The maxillary with its well-worn P^4/M^1 teeth fits in its sizes to the Late Pleistocene *C. l. spelaeus* skulls from various sites in Europe [7, 11 - 13], which sexual cranial dimorphism (larger males, smaller females) is not yet well studied. The maxillary (Fig. **3.1**) has strongly worn breaking scissor P^4/M^1 teeth and comes from an elderly individual. There are few descriptions and figurations of longbones, such as material from the German Sophie's Cave or the Devils Chamber Cave [13, 14]. Therefore, unpublished longbones from several German caves were used, which are currently under further study. The proportions of the Hermann's Cave wolf femorae are quite different especially in their lengths (Fig. **2.11** length: 285 mm, distal width: 49 mm; Fig. **2.13** length: 257 mm, distal width: 47 mm). The nearly 3 cm longer femur is from a male, which similar lengths are found in femora from the Zoolithen Cave or the Seweckenberge.

The much shorter female femorae are also present in the Zoolithen Cave. The half radius and ulna were directly compared to similar proportioned smaller female material from the Martins Cave, Perick Caves, Zoolithen Cave, and the figured radius of the Devils Chamber Cave [14]. Possibly, those and other material figured on the plate (Fig. **2**) belongs to a single female individual from the Bear Hall cave part, including the complete tibia (Fig. **2.16** female - length: 236 mm, distal width: 30 mm). Compared to tibiae from the Zoolithen Cave and the Keppler Cave, this falls into smaller lengths in females.

The MNI of 3 individuals can be separated based on the femorae to one male, one female and one cub. The incomplete tibia with its expanded shaft and deformed distal joint (Fig. **2.17**) is of unclear pathology. Three *C. l. spelaeus* longbones, both subadult femorae (cracked) and one tibia (proximally chewed) expose carnivore bite and crack damage (Figs. **2.14, 2.25, 2.17**). The medium-large bite mark sizes (around 3 mm in width) and chewed joint damages indicate the wolf itself (or even only fox) as producer, compared to modern [*e.g.* 15 - 16] and extinct [17 - 19] wolf, hyena or lion guilt longbone damages.

Ice Age Spotted Hyena Remains from the Hermann's Cave

The eight *Crocuta crocuta spelaea* (Goldfuss, 1823) (former "cave hyena") [1] bones figured herein (Fig. **3**) were excavated historically and only in the first ten meters from the former Hermann's Cave entrance [7]. The cranial material consists only of a single subadult individual maxillary with the P^3 (Fig. **3.1**), which fits well in comparison with the European cranial record [2], or *e.g.* the Czech Srbsko Komin Cave cranial material with similar P^4 morphotypes [20]. Also a femur shaft with modern distal damage belongs to a subadult, which are similarly figured (Fig. **3.2**) for hyena cubs from the Srbsko Komin Cave [20]. All other pedal bones of the fore limb (Figs. **3.2**) and hind limb (Fig. **3.2**), are from subadult-elderly individuals and are well to compare to material from *e.g.* the Srbsko Komin Cave [20]. The calcaneus is pathologic with bone growth on its surface (exostoses), which most probably must have resulted from a trauma of intraspecies fights, resulting from hyenas targeting the hind limb with their bites [21].

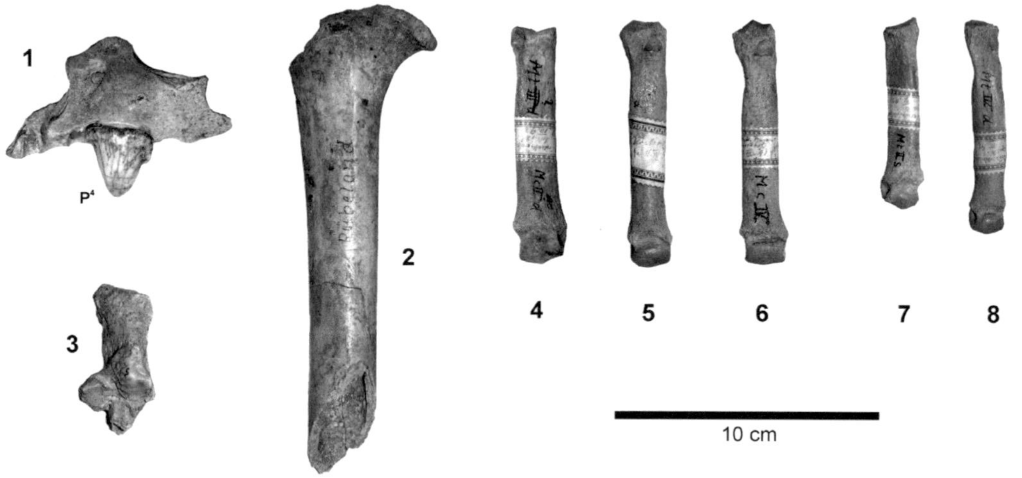

Figure 3. Late Pleistocene spotted hyena *Crocuta crocuta spelaea* (Goldfuss, 1823) remains (most probably a single subadult individual) from the Bear Hall of the Hermann's Cave, Rübeland, Saxony-Anhalt (Germany). 1. Left maxillary with P^4 (MLUIFG), lateral. 2. Left parthologic (exostoses) calcaneus (MLUIFG), cranial. 3. Right femur shaft of a subadult (LMNB), cranial. 4. Right Mc II (LMNB), dorsal. 5. Right Mc IV (LMNB), dorsal. 6. Right Mc IV (LMNB), dorsal. 7. Left Mt II (LMNB), dorsal. 8. Right Mt IV (LMNB), dorsal (composed from [7]).

Dating of the Hyena and Wolf Bones

The details can be found in the Chapter Geology. The wolf remains were found in the autochthonous Late Pleistocene cave sediments with red cave clay, or on the surface between cave bear bones being covered by speleothems, breccia and fallen limestone blocks [7]. The youngest speleothem generation and candle stalagmites covered the *Canis lupus spelaeus*, cave bear or lion bones and archaeological Aurignacian finds in the Hermann's Cave [7]. It's covering speleothem layer date all those remains including the wolves, to be older than 23.130 ± 711 and 24.260 ± 568 BP [7], and are therefore do not belong to European or arctic wolves from the later LGM period. The wolf bones can be dated indirectly using the "cave bear clock" (see Chapter Cave Bears), between approximately 130.000-24.000 BP into the Late Pleistocene glacial (= MIS 3-5d).

Hyena or Wolf Den and Guilt

Obviously, the main hyena/wolf den area is not yet excavated, and is still under LGM-frost breccia coverage within the first 0-10 meters distance from the former entrance (cf. Fig. **1**). This might explain also the low numbers of the carnivore and their guilt remains. The non-cave bear material is limited with only 46 bones (Table **1**). However, the large amount of wolf/hyena remains are typical at extinct Late Pleistocene [*e.g.* 22, 23] and extant [*e.g.* 24, 25] wolf or spotted hyena dens.

The Ice Age wolf *C. l. spelaeus* dominates with 43% the NISP (Fig. **1**) with a ratio of one male/one female and one cub. Compared to a modern long-term used Canadian wolf cave den in the boreal forest Rocky Mountains, such a population structure correlates to a wolf pack at a communal den [26]. There is only this single short description available that reports on imported megafauna body parts, which were further chewed by the adults and cubs, leaving in the first ten meters of the smaller cave a bone accumulation [26]. Detail studies about wolf caused bone accumulations at their dens are still missing, which makes the comparison to Late Pleistocene wolf (or wolverine) bone assemblages still problematic. The main megafauna guilt for the Canadian wolves is the reindeer, followed by musk ox or forest bison [*e.g.* 26 - 28]. The Late Pleistocene wolf guilt fauna assemblage at the Hermann's Cave is dominated by *Rangifer tarandus*, which is still in general the favorite guilt of modern Canadian or Scandinavian arctic wolves [26 - 28]. It is much lesser represented in amounts at Late Pleistocene Ice Age spotted hyena den sites, where mainly only chewed antler bases of dropped antlers were found, that have been collected by hyenas [*e.g.* 2, 4]. The high amount of wolf, reindeer (22%) such as alpine medium-sized *Rupicapra rupicapra* (5%) and few amounts of mammoth/rhinoceros bones (Fig. **4**) indicates a wolf den, most probably a commuting den.

This seems to have been used for a very short time period by hyenas as a den site, which is more typical at larger cave bear den caves in Europe [2, 13, 19]. The Ice Age spotted hyena *C. c. spelaea* material is from at least one subadult (MNI = 1). Maybe it comes from different individuals. At Late Pleistocene European spotted hyena den sites, especially communal dens, the amount of hyenas counts range between 25-40% [2], whereas at natal/birth dens, this amount decreases down to 10-15% [2]. At the Hermann's Cave, the 17% hyena remains would fall into the latter range, also fitting would be the remains of a possibly single subadult individual. If the cave was used as a hyena natal den, then for a very short period/season, which explains also the low amount of imported guilt remains. To those, mammoth and woolly rhinoceros similar damaged bones belong to [*e.g.* 2, 8, 14], which are represented at the Hermann's Cave few (= 2% and 7% of NISP) (Table **1**).

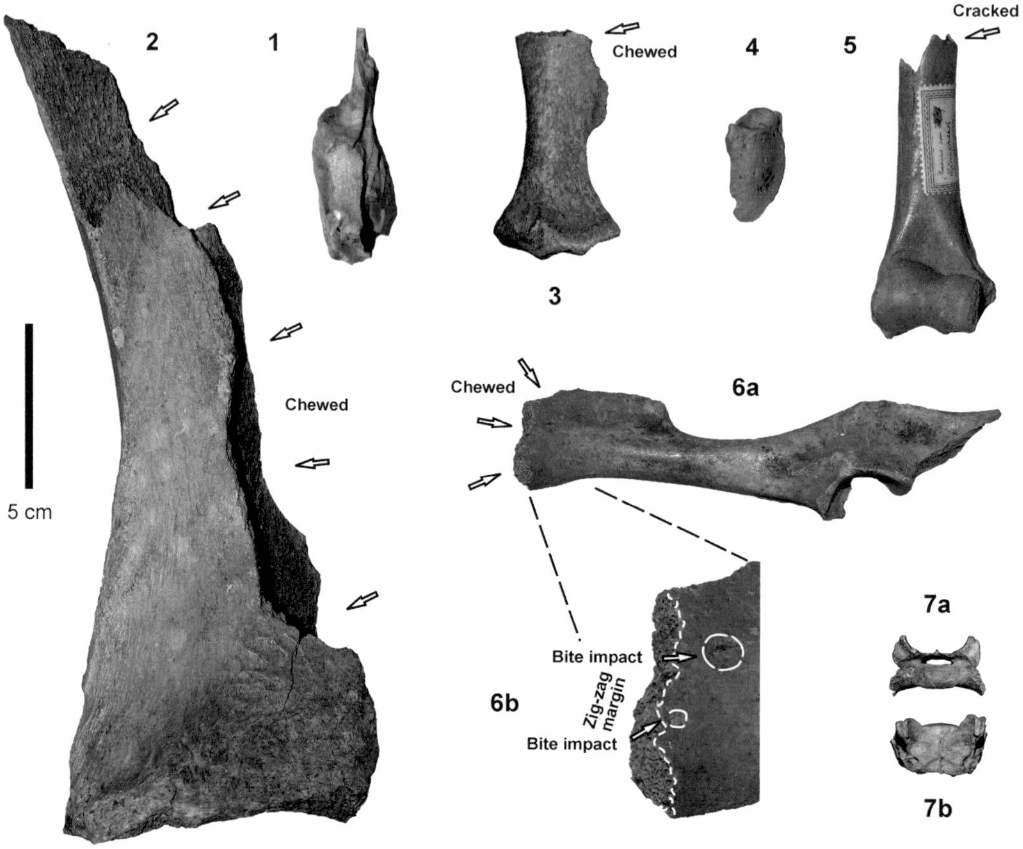

Figure 4. Late Pleistocene Ice Age spotted hyena and Ice Age wolf imported guilt remains anterior Hermann's Cave part. 1. Woolly rhinoceros *Coelodonta antiquitatis* upper jaw molar tooth fragment (LDA). 2. Woolly rhinoceros *Coelodonta antiquitatis* scapula with chew damage (LMNB), lateral. 3. Woolly rhinoceros *Coelodonta antiquitatis* neonate or just born cub femur with proximally chewed part (HC), cranial. 4. Reindeer *Rangifer tarandus* phalanx III (LDA), dorsal. 5. Chamois *Rupicapra rupicapra* half distal cracked humerus (LMNB), caudal. 6. Chamios *Rupicapra rupicapra* right coxa with marginal wolf-fox chew and small bite mark damage on the ischium (LMNB), lateral (a) and detail (b). 7. Red fox *Vulpes ulpes* sacrum (HC), cranial (a) and dorsal (b).

The main wolf den sites in the Harz region are the Hermann's Cave as described herein, the Seweckenberge and Westeregeln, which all three are along the Bode River that passed the Hermann's Cave (Fig. **5**). Other hyena and wolf dens are found similarly along river valleys/terraces [9]. Extant northern hemisphere wolves chose their dens along caribou migration trails following up to boreal forest zones [29, 30]. The Late Pleistocene wolf packs in northern Germany seem to have used seasonal den sites, especially by following reindeer herds during migration to their summer feeding/birth places in European boreal forests [31] including the boreal forest Harz Mountains (Fig. **5**). The Late Pleistocene wolf dens seem to overlap in most cases the spotted hyena dens (Fig. **1**, **5**), which were also chosen obviously along the big game migratory channels causing competition between both pack/clan hunters.

Table 1. Late Pleistocene Ice Age bone megafauna bone assemblage (cave bears excluded), and some important metrics of complete bones from the first 20 m of the anterior Hermann's Cave (= wolf, and sporadic hyena den, cf. Fig. 4).

Species	Bones	NISP
Mammuthus primigenius	Small fragments of tusk	1
Coelodonta antiquitatis	M-tooth fragment, scapula and cub femur	3
Rupicapra rupicapra	Half distal humerus and coxa	2
Crocuta crocuta spelaea	Maxillary, femur shaft, 4 metapods and calcaneus	8
Canis lupus spelaeus	Maxillary, I-tooth, scapula, half proximal radius, incomplete ulna, right Mc III, coxa and ischium, C2, C3 vertebra, lumbar vertebra, calcaneus, astragalus, 2 complete (male - length: 285 mm, distal width: 49 mm; female - length: 257 mm, distal width: 47 mm) and one half distal femur, 2 x cracked cub femur shafts, one complete (female - length: 236 mm, distal width: 30 mm) and one distal half tibia, right Mt IV	22
Vulpes vulpes	Sacrum	1

Cave bears were guilt for both, hyenas and wolves, in boreal forests over European middle high elevated mountain regions [2, 13, 18, 19]. The larger amount of large carnivore damaged cave bear bones from the Hermann's Cave (see Chapter cave bears) is typical at all European cave bear dens [13, 18, 19] and a result of cave bear scavenging in the easy accessible horizontal caves, especially during winter (hibernation) and spring (cub raising) times [13, 17, 19]. The problem of exact NISP statistics related to lion, hyena or wolf cave bear bone damage is difficult because damage patterns can be the same in all three predator activities (*e.g.* chewing of soft joints and leaving shafts [13, 15, 17 - 19]). Only puncturing (oval-round holes in long bone compacta only) and cracking of the most massive bone shafts of adult cave bears can be attributed securely to hyenas and their breaking scissor premolar teeth, which produce the typical pseudo-bone flutes [17] (see Chapter cave bears).

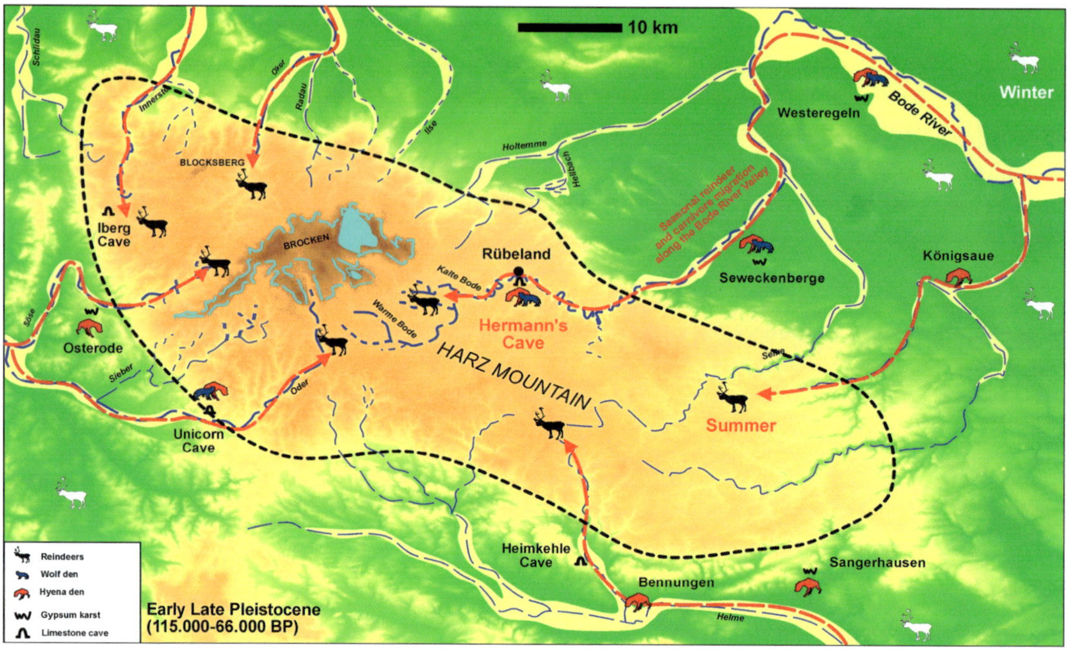

Figure 5. Position of Ice Age spotted hyena *Crocuta crocuta spelaea* and wolf *Canis lupus spelaeus* dens along the river valleys and seasonal megafauna (especially reindeer) migration routes. Reindeers and other megafauna migrated in spring/summer into the mountains.

REFERENCES

[1] Goldfuss GA. Osteologische Beiträge zur Kenntnis verschiedener Säugethiere der Vorwelt. VI. Ueber die Hölen-Hyäne (*Hyaena spelaea*). Nov Act Phys-Med Acad Caes Leopold-Carol Nat Curios 1823; 3(2): 456-90.

[2] Diedrich C. Palaeopopulations of Late Pleistocene top predators in Europe: Ice Age spotted hyenas and steppe lions in battle and competition about prey. Paleont J 2014; pp. 1-34.

[3] Diedrich C. The Late Pleistocene *Crocuta crocuta spelaea* (Goldfuss, 1823) population from the Emscher River terrace hyena open air den Bottrop and other sites in NW-Germany – woolly rhinoceros scavengers and their bone accumulations along rivers in lowland mammoth steppe environments. Quat Int 2012; 276/277: 93-119. [http://dx.doi.org/10.1016/j.quaint.2011.07.046]

[4] Diedrich C. Recycling of badger/fox burrows in Late Pleistocene loess by hyenas at the den site Bad Wildungen-Biedensteg (NW, Germany): woolly rhinoceros killers and scavengers in a mammoth steppe environment of europe. J Geol Res 2013; 2013: 1-31. [http://dx.doi.org/10.1155/2013/190795]

[5] Leonard JA, Vilà C, Fox-Dobbs K, Koch PL, Wayne RK, Van Valkenburgh B. Megafaunal extinctions and the disappearance of a specialized wolf ecomorph. Curr Biol 2007; 17(13): 1146-50. [http://dx.doi.org/10.1016/j.cub.2007.05.072] [PMID: 17583509]

[6] Diedrich C. Holotype skulls, stratigraphy, bone taphonomy and excavation history in the Zoolithen Cave and new theory about Esper's "great deluge". Quat Sci J 2014; 63(1): 78-98.

[7] Diedrich C. Evolution, Horste, Taphonomie und Prädatoren der Rübeländer Höhlenbären, Harz (Norddeutschland). Mitt Verb dt Höhlen- Karstf 2013; 59(1): 4-29.

[8] Diedrich C. Late Pleistocene *Crocuta crocuta spelaea* (Goldfuss, 1823) clans as prezewalski horse hunters and woolly rhinoceros scavengers at the open air commuting den and contemporary Neanderthal camp site Westeregeln (central Germany). J Archaeol Sci 2012; 39(6): 1749-67. [http://dx.doi.org/10.1016/j.jas.2012.01.013]

[9] Diedrich C. Seltene Freilandfunde der späteiszeitlichen Fleckenhyäne *Crocuta crocuta spelaea* (Goldfuss 1823) in Sachsen-Anhalt und Beitrag zu Horsttypen der letzten Hyänen im Jung-Pleistozän von Mitteldeutschland. Z Mitteldt Vorgesch 2014; 94: 1-19.

[10] Schütt G. Die jungpleistozäne Fauna der Höhlen bei Rübeland im Harz. Quartär 1969; 20: 79-125.

[11] Reynolds SH. Monograph of the British Pleistocene Mammalia II(III), the Canidae. Monogr Pal Soc London 1909; pp. 1-28.

[12] Germonpré M, Sablin MV, Stevens RE, *et al.* Fossil dogs and wolves from Palaeolithic sites in Belgium, the Ukraine and Russia: Osteometry, ancient DNA and stable isotopes. J Archaeol Sci 2009; 36(2): 473. [http://dx.doi.org/10.1016/j.jas.2008.09.033]

[13] Diedrich C. Late Ice Age wolves as cave bear scavengers in the Sophie's Cave of Germany – extinctions of cave bears as result of climate/habitat change and large carnivore predation stress in Europe. ISRN Zool 2013; 1-25. [http://dx.doi.org/10.1155/2013/138319]

[14] Diedrich C. The *Crocuta crocuta spelaea* (Goldfuss, 1823) population and its prey from the Late Pleistocene Teufelskammer Cave hyena den besides the famous Palaeolithic Neanderthal Cave (NRW, NW Germany). Hist Biol 2011; 23(2/3): 237-70.

[15] Domínguez-Rodrigo M, Gidna AO, Yravedra J, Musiba C. A Comparative Neo-Taphonomic Study of Felids, Hyaenids and Canids: an Analogical Framework Based on Long Bone Modification Patterns. J Taphonomy 2012; 10(3/4): 147-64.

[16] Nohemi S, Arsuaga JL, Haynes G. Taphonomic comparison of bone modifications caused by wild and captive wolves (*Canis lupus*). Quat Int 2014; 330: 126-35. [http://dx.doi.org/10.1016/j.quaint.2013.08.017]

[17] Diedrich CG. 'Neanderthal bone flutes': simply products of Ice Age spotted hyena scavenging activities on cave bear cubs in European cave bear dens. R Soc Open Sci 2015; 2(4): 140022. [http://dx.doi.org/10.1098/rsos.140022] [PMID: 26064624]

[18] Diedrich C. Cracking and nibbling marks as indicators for the Upper Pleistocene spotted hyaena as a scavenger of cave bear (*Ursus spelaeus* Rosenmüller, 1794) carcasses in the Perick Caves den of Northwest Germany. Abh Naturhist Ges Nürnberg 2005; 45: 73-90.

[19] Diedrich C. Cave bear killers and scavengers from the last Ice Age of central Europe: Feeding specializations in response to the absence of mammoth steppe fauna from mountainous regions. Quat Int 2011; 255: 59-78. [http://dx.doi.org/10.1016/j.quaint.2011.06.048]

[20] Diedrich C. The largest Late Pleistocene hyena population from the Srbsko Chlum-Komín Cave (Czech Republic) and its prey in a commuting and prey depot cave den of Central Europe. Acta Zool (in press).

[21] Diedrich C. Late Pleistocene European Ice Age spotted hyena aging and battle palaeopathologies. Int. J. Paleopathol.

[22] Stiner MC. Comparative ecology and taphonomy of spotted hyenas, humans, and wolves in Pleistocene Italy. Rev Paleobiol 2004; 23(2): 771-85.

[23] Diedrich C. Late Pleistocene spotted hyena den sites and specialized rhinoceros scavengers in the Thuringian Mountain Zechstein karst (Central Germany). Quat Sci J 2015; 64(1): 29-45.

[24] Ballard WB, Dau JR. Characteristics of gray wolf, *Canis lupus*, den and rendezvous sites in

southcentral Alaska. Can Field Nat 1983; 97: 299-302.

[25] East ML, Hofer H, Türk A. Functions of birth dens in spotted hyaenas (*Crocuta crocuta*). J Zool (Lond) 1989; 219: 690-7.
[http://dx.doi.org/10.1111/j.1469-7998.1989.tb02614.x]

[26] Mech LD, Packard JM. Possible use of (*Canis lupus*) den over several centuries. Can Field Nat 1990; 104: 484-5.

[27] Hansen IJ, Johnson CJ, Cluff HD. Synchronicity of movement paths of barren-ground caribou and tundra wolves. Polar Biol 2013; 36: 1363-71.
[http://dx.doi.org/10.1007/s00300-013-1356-y]

[28] Klaczek MR, Johnson CJ, Cluff HD. Den site selection of wolves (*Canis lupus*) in response to declining caribou (*Rangifer tarandus groenlandicus*) density in the central Canadian Arctic. Polar Biol 2007; 38: 2007-19.
[http://dx.doi.org/10.1007/s00300-015-1759-z]

[29] Kaartinen S, Luoto M, Kojola I. Selection of den sites by wolves in boreal forests in Finland. J Zool (Lond) 2010; 281: 99-104.
[http://dx.doi.org/10.1111/j.1469-7998.2009.00678.x]

[30] Heard DC, Williams TM. Distribution of wolf dens on migratory caribou ranges in the Northwest Territories, Canada. Can J Zool 1992; 70: 1504-10.
[http://dx.doi.org/10.1139/z92-207]

[31] Weinstock J. Late Pleistocene reindeer populations in Middle and Western Europe. Bio Archaeologica 2000; 3: 1-307.

CHAPTER 6

Last Glaciation Maximum (LGM)

Abstract: Four main ponor levels were formed from the Bode River valley terrace sided within 39 meters elevation distances, each about 10 meters deeper by the subsurface drainage. This correlates to Bode River valley glaciers and melting waters during the Elsterian, Saalian and Weichselian glacials, and finally to the modern Holocene groundwater level. The active underground river transports Mt. Brocken peak granite and Bode upstream Palaeozoic rock type gravels into the lowermost ponor level. In the Early-Middle Late Pleistocene (MIS 5-3) there was a change between few colder and warmer short-time periods (MIS 5a-d). In the colder periods, the Mt. Brocken peak must have started to built up an Ice cap in elevations of 1.200 meters. This is low, compared to other European regions, but the Harz was at that time already close to the south moving Scandinavian Inland Glacier. The boreal forest conditions were perfect for cave bears, but also Neanderthal human groups. About 77.0000 years ago, the MIS 4 cold period must have increased the glacial conditions, which found a peak in the Last Glacial Maximum (= LGM) 19.000 years ago. At that time of the LGM, the Scandinavian Inland Ice shield reached even Hamburg and Berlin regions, which cold winds formed in northern Germany and the Harz foreland lowlands a periglacial tundra landscape with frost polygons. Already around 26.000 years ago, the conditions were even too cold for the mammoth steppe magafauna and cave bears, which disappeared together with humans in complete northern Europe. The Bode Valley Glacier finally resulted in the Hermann's Cave entrance and deeper ceiling collapses, which frost breccias closed the former entrance completely. Three caves/cavities at Rübeland in the Bode Valley, Harz Mountains (northern Germany) contain bone assemblages that resulted from carnivore mammal wolverine *Gulos gulo*, polar fox *Vulpes alopex* and mustelid activities such as large carnivore eagle owl *Bubo bubo* subsp. bird den. All remains refer to an arctic tundra/taiga fauna that is mixed with an alpine rocky and cave-rich area which supports the LGM glaciation theory of the Brocken Peak around 19.000 BP (= Last Glacial Maximum). The valley glaciers and Mt. Brocken Ice filed (and Scandinavian shield) disappeared at the end of the MIS2 between 16-12.000 years ago.

Keywords: Late Pleistocene (MIS 2-5d), Last Glacial Maximum (= LGM), landscape and glacier development, River terrace and cave ponor level correlation, Glacial/alpine fauna, Predator cave dens along Bode River valley, Harz Mountains, Northern Germany.

Cajus G. Diedrich

GLACIAL SIGNS AND LANDSCAPE OUTSIDE THE HERMANN'S CAVE

Initial glaciomorphology studies in the Harz Mountains distinguished only two glaciation periods in the Harz Mountains [1 - 5], but three in its fore lands by fluvial/glacial sediments [6 - 12]. Outside the caves, the geomorphology in higher elevations close to the Mt. Brocken allowed recently distinguishing between three different elevated glacier periods in different Marine Isotope Stages (= MIS) [age after 13], which seem to represent the highest Elsterian Glacial (MIS 10, 364.000-400.000 BP [age after 14]), medium elevated late Saalian Glacial (MIS 6, 130.000-190.000 BP [age after 14]) and lowermost Weichselian Glacial (MIS 2, Last Glacial Maximum = LGM [15]: around 19.000 BP [age after 14]) (Fig. **1A**). Typical geomorphology shapes left by those glaciers are U-shaped valleys upstream of the Bode (Kalte Bode branch) below the Mt. Brocken Peak above the village Elend (Fig. **1B-C**), and north of the village Königshütte (Figs. **1E-G**, **2**) [13]. Elsterian Glacial deposits are known only from river terraces on the outer slopes of the Harz Mountain and its forelands [6 - 12]. An ice cap seems to have covered, similar to the Saalian Glacial, the Harz Mt. Brocken peak [13] (Fig. **1A**) while the inland ice reached at that time even the northern slope of the Harz [2, 16]. The Bode Valley glaciers must have passed, at that time the second time, all larger caves: the Unicorn (by Oder Valley Glacier), the Baumann's and Hermann's Caves (by Bode Valley Glacier), which explains the beginning of the first larger ponor cave systems that development in the region during the Elsterian Glacial [13] (see Chapter Geology). In the Weichselian Glacial MIS 5b and 5d stadial oscillations the development of a first glacier cap must have developed resulting in valley glaciers [13]. The ice shield must have started to grow in the Mt. Brocken peak during MIS 4 increasing climate temperature drop starting about 77.000 BP [13]. The growing ice cap and valley glacier melting waters explain first an increasing of the Bode River terrace elevation.

19.000 BP - Bode River Valley LGM Glaciers

The ice field or ice cap of the Mt. Brocken reached its maximum extension and thickness on the Brocken peak about 19.000 years ago (MIS 2, LGM: Fig. **1A**) [13], such as it is globally recorded [15]. This, and the valley glaciers, are comparable to *e.g.* the Rocky Mountain Columbian Ice field [17]. In some valleys, or kar areas, three stages of the LGM glaciers can be reconstructed on arch-like valley end moraines, which are partly preserved (Figs. **1E-G**) [13]. The snow accumulated mainly on the SE slope of the Mt. Brocken peak which resulted in some valley glaciers as result of the NW-Atlantic climate and SE wind direction impact [5, 13]. Those are best mapped for the Oder Valley [1, 13]. In the Bode Valley upstream the signs are clearer, but in the region of the village

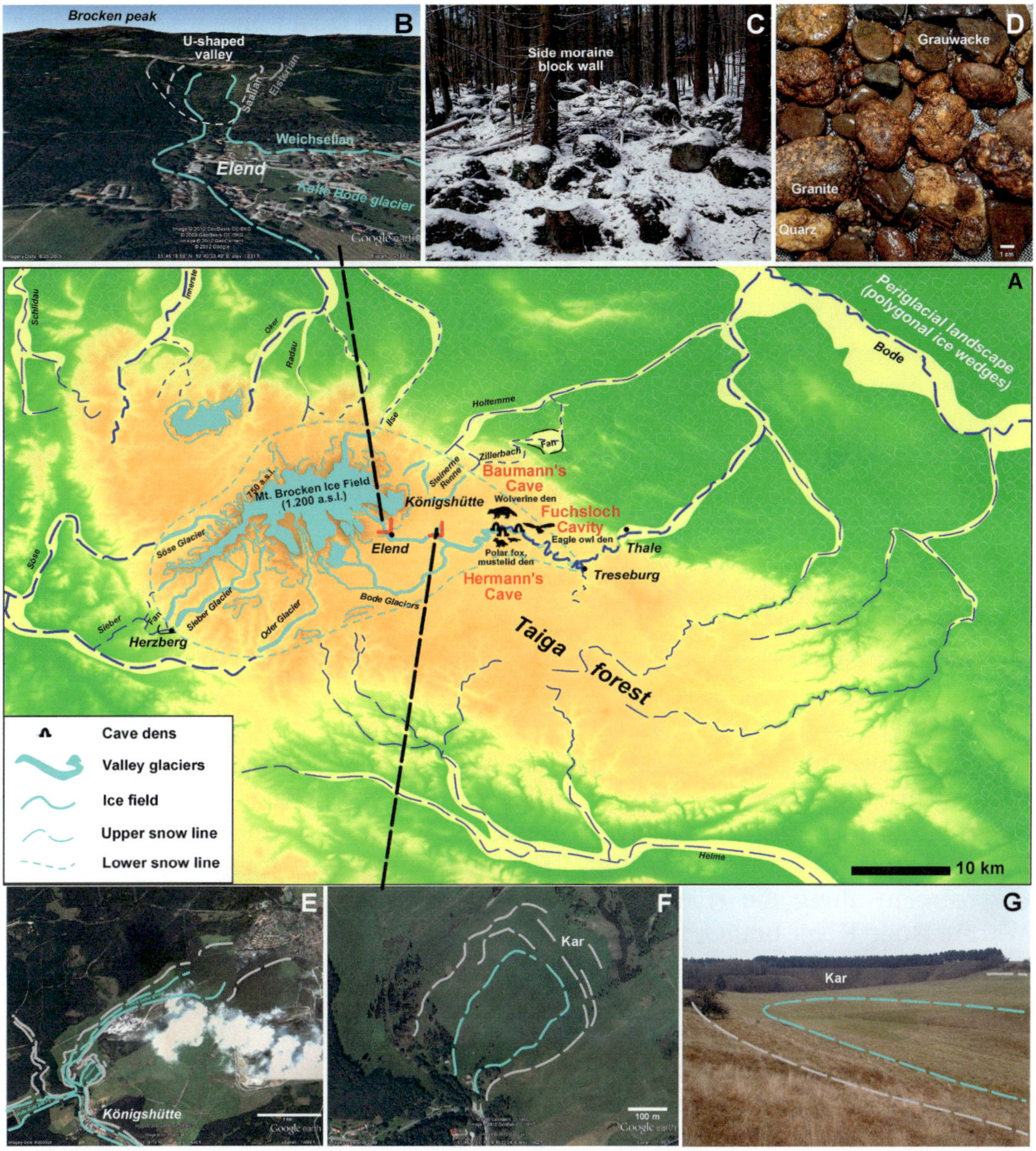

Figure 1. A. The Harz Mountain (northern Germany) Brocken Ice cap and valley glaciers in their maximum extension during the Late Pleistocene MIS 2 (Last Glacial Maximum = LGM), taiga and tundra landscapes and predator den sites along the Bode Valley near Rübeland. B. The Kalte Bode Valley Glacier elevations at Elend village. In the valley several typical glacier structures occur very similar to the Oder Valley Glacier. C. Side moraine structure walls built of rounded granite blocks. D. The Kalte Bode river gravels dominate at Elend with the Mt. Brocken granite, few quarz and Kulm Grauwacken (composed from [6] and modified).

Rübeland, it remains unclear, if the LGM glacier really passed the Hermann's Cave in its maximum extension as suggested [13]. At this coldest Weichselian

Glacial period, periglacial permafrost landscapes with frost polygons were present in northern Germany (Figs. **1A**), the northern Harz lowlands [13]. This resulted in the Scandinavian Inland Ice Shield, which had its maximum southern extension down to Hamburg-Berlin, only 150 km far from the Harz Mountain [2, 5, 8]. Its southeastern directed arctic glacier winds caused a taiga-like landscape in the northern Harz lowlands [2, 5, 8, 13], which replaced already before the LGM about 24.000 BP the mammoth steppe vegetation, whereas the forest in the outer Harz slopes are expected to have been taiga forests [13].

The Kalte Bode River branch starts below the Brocken Mountain where a deep eroded U-shaped valley above the village Elend is one of the best examples and highest elevated U-shaped valley found in the Harz (Figs. **1A, B**). Three elevations are developed with the lowermost being of Weichselian aged glacier lateral moraine deposits (Figs. **1C**) starting on elevations of Lat. 51°45'7.40"N, 10°40'44.13"E, 615 a.s.l [13]. Some kilometers downstream, close to the village Königshütte, two different elevated lateral moraines are visible on both sides (Lat. 51°44'58.96"N, Long. 10°44'52.83"E, 469 a.s.l., Fig. **2A**) [13]. Those are about 10 km upstream from the Hermann's Cave. At Königshütte, three valley glaciers were fusing (Fig. **2E**) [13]. Northwest of Königshütte and besides the Teich Valley, a kar depression is present, such as a short glacier, whereas lateral morains can be mapped in two elevations which might represent different Weichselain retreat stages or a reused former Saalian Ice age kar (Figs. **1F-G**) [13]. Water corrosion into volcanite rocks by glacier surface water is documented there on a rock wall with concave erosive reliefs (Fig. **2C**) [13]. Hundred meters downstream from the end moraine arches, the volcanites are polished to convex morphologies by ice abrasion (Fig. **2E**). Such abrasion is found one km further downstream along the river valley at Middle Devonian limestones [13]. The Warme Bode River branch terminal and lateral moraine walls were found parallel to the Kalte Bode valley eight km upstream of the Hermann's Cave nearby Bodfeld (Lat. 51°44'14.77"N, Long. 10°45'50.18"E, 431 a.s.l.) [13]. More upstream, at Tanne, a terminal moraine wall is exposed on the field, and in front, a small dead ice block depression [13]. The walls built an arch at the end. Further upstream, where the Warme Bode is branching (Lat. 51°41'53.71"N, Long. 10°40'18.85"E, 493 a.s.l.), Saalian and Weichselian lateral moraines seem to be preserved, which have different elevations (Saalian older ones are several meters higher).

The Bode Valley Glacier must have fused with the two main branches below Königshütte, and seem to have passed the Rübeland caves [13] (Fig. **1A**). Kames/lateral moraine deposits are identified as relic sediments only in the former entrance and western part of the Baumann's Cave [13], whereas those might be of MIS 6 in age. Those few rounded unsorted limestone blocks and gravels contain

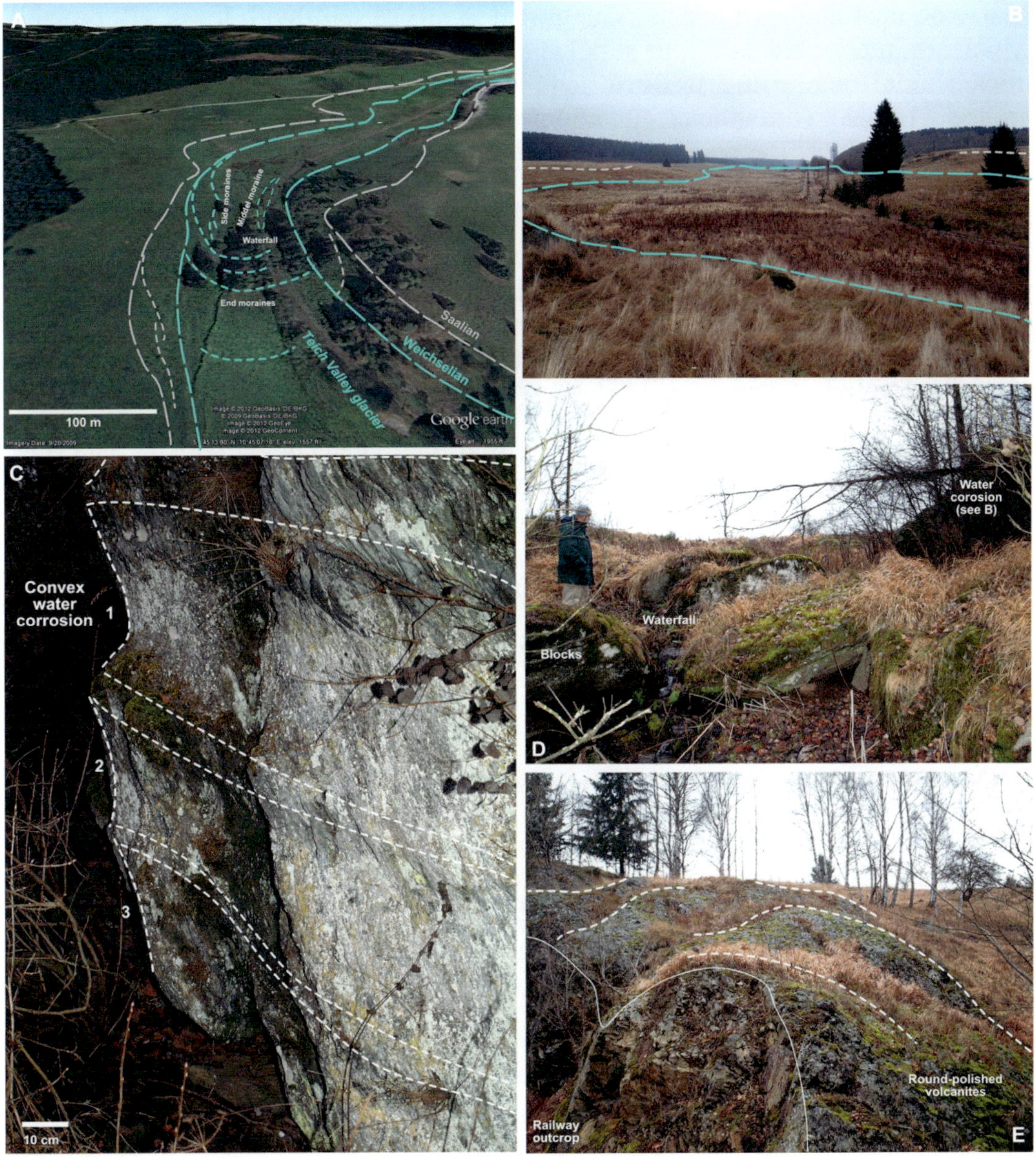

Figure 2. The Late Pleistocene Teich Valley Glacier, a side valley glacier of the Kalte Bode Glacier above Königshütte (Harz Mountain, northern Germany). A. Map with reconstruction outlines of Saalian higher elevated and Weichselian lower elevated valley glaciers. B. View into the U-shaped valley. In the area of the waterfall several typical structures occur, such as C. Convex water corroded (several elevations) hard-rock volcanites, D. Water fall within endmoraine blocks, E. By ice round-polished hard-rock volcanites. All structures are found in the waterfall area, which seem to have some end moraine and a middle moraine wall (from [13]).

nearly no upstream extraclasts. The gravel is mixed with with loess, silt or sand, which are interpreted as till, or subglacial outwash [13]. The maximal two meters thick fluvial coarsening upward sequence was washed from an elevation of around 413 a.s.l. (the glacier elevation/margin) into/within the Baumann's Cave during glacier surface water floods [13]. Sediment eroded from kames-terraces and lateral moraines which sediments must have been washed into the cave area in short-term events including reworking of cave sediments/bones/artifacts, when the glacier (estimated by the cave entrance elevation and deposits about 35 meters thick) passed the valley at Rübeland [13]. Most probably its last terminal moraine (maybe also Saalian in age) ended in the village Treseburg, several kilometers from the slope to Thale, but from Treseburg, the valley is V-shaped and not glacially formed at any time [13].

The MIS 2/1 climate changing times revered the glacier development. The Harz Mountain glaciers melted finally completely between 14-12.000 BP and left glacier valley landscape geomorphologies on the southeastern Mt Brocken slope valleys [13] (Fig. **1A**). Within the changing interstadial/stadials the Bode River must have started to erode in elevation around 386 a.s.l. at Rübeland and in front of the Hermann's Cave rapidly (less than 10.000 years) by the retreating glaciers/ice field melting waters [13]. The Hermann's Cave entrance was closed completely in the Post-LGM by further frost breccias (Figs. **3A-B**), whereas the ceiling collapses seem to fall mainly into this period [13].

LGM and Meiendorf-younger Dryas Predator Dens along the Bode Valley

Three caves/cavities in the Bode Valley nearby Rübeland contain faunal remains, which bone assemblages resulted from carnivore mammal wolverine, polar fox and mustelid activities such as large eagle owl [13]. All remains refer to an arctic tundra/taiga fauna that is mixed with alpine rocky and cave-rich elements (Table **1**) [13].

In the Hermann's Cave former entrance area (first 10-15 meters) a polar fox/sable den was present (Figs. **1**, **3**) [13]. A half *Vulpes lagopus* skull (Fig. **1.1**) and postcranial elements are the most abundant carnivore remains (NISP = 9%). A *Mustela* cf. *erminea* skull (Figs. **4.2**) [18, 22] and jaws and few postcranial bones represent the small mustelid carnivores (NISP = 1%) [13].

The fox and mustelid bone assemblage overlap (Table **1**), but mainly in the micromammals, whereas sievings for the microfauna in the Hermann's Cave fox/mustelid den area are missing. Only few remains are present of the lemmings *Lemmus lemmus*, *Dicrostonyx torquatus* and vole *Microtus oeconomus* (Table **1**) fit to *V. lagopus* but also *M. nivalis* guilt spectrum [21, 22]. The snow (*Lepus timidus*, Fig. **4.3**) and European hare (*Lepus europaeus*, Fig. **4.4**) remains (NISP =

12%, Fig. **3**) are typical for polar fox guilt [20, 21]. The main and larger prey remains seem to be accumulated by polar fox. Larger mammals are few represented only with some Przewalski horse (*Equus ferus przewalskii*, NISP = 4%) leg (Fig. **4.5**) and thorax (costae) remains, whereas some bone flakes seem to belong to alpine elements (*Rupicapra, Capra*). To the polar fox food spectrum belong the birds of the western capercaillie *Tetrago urogalus* NISP = 42%) and willow grouse *Lagopus lagopus* [20, 21] (NISP = 31%) which count together 73% of all bones (Fig. **3**).

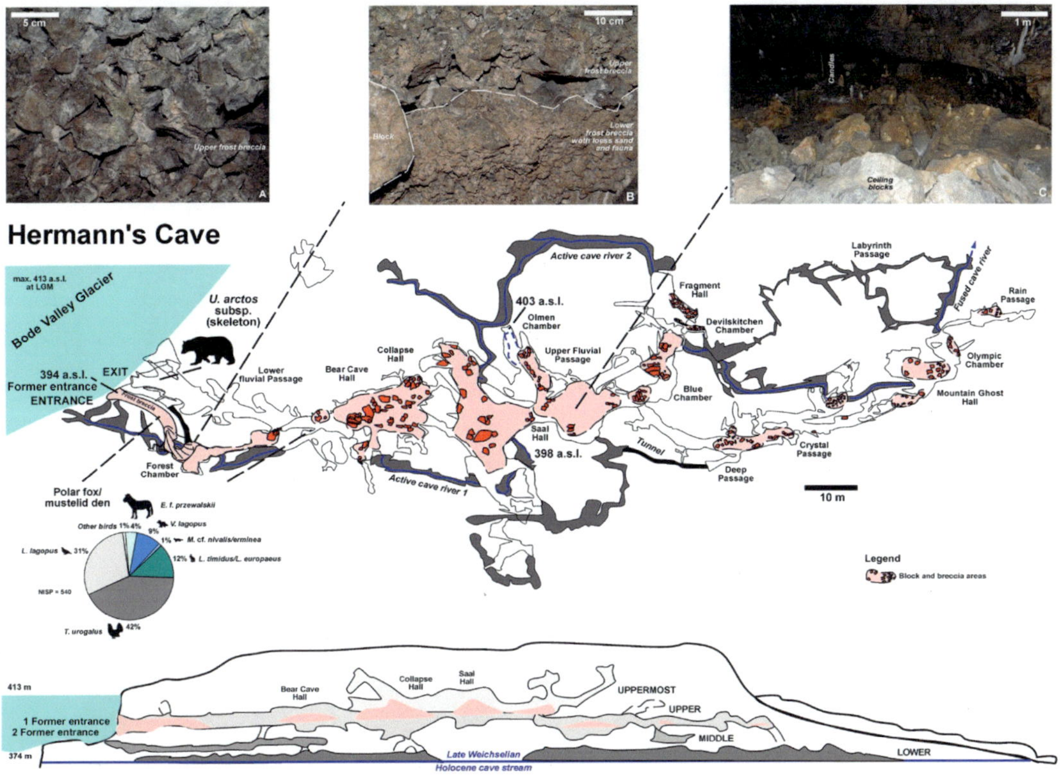

Figure 3. Hermann's Cave during the LGM (MIS 2/1 boundary). Arctic fox and mustelid dens in the former entrance area (LGM frost breccias) and main ceiling collapse blocks (red) with more thick frost breccias areas (reddisch). A. Upper layer of frost breccia in former entrance without fauna. B. Boundary Upper and Lower frost breccia with loess sand and arctic/alpine fauna. C. Large blocks (block-field) sloping into the Sall Hall. Very small candles on the blocks are Post-LGM in ages (updated in the fauna assemblage and modified from [13]).

Above the Baumanns's Cave (Fig. **1**) a sinkhole was used by wolverines (*Gulo gulo*) and polar fox (*Vulpes lagopus*) as den [13]. The polar fox den position on the plateau margin at the crests of the slope in a loess refilled doline which was partly and seasonally (summer/winter) under permafrost conditions along the

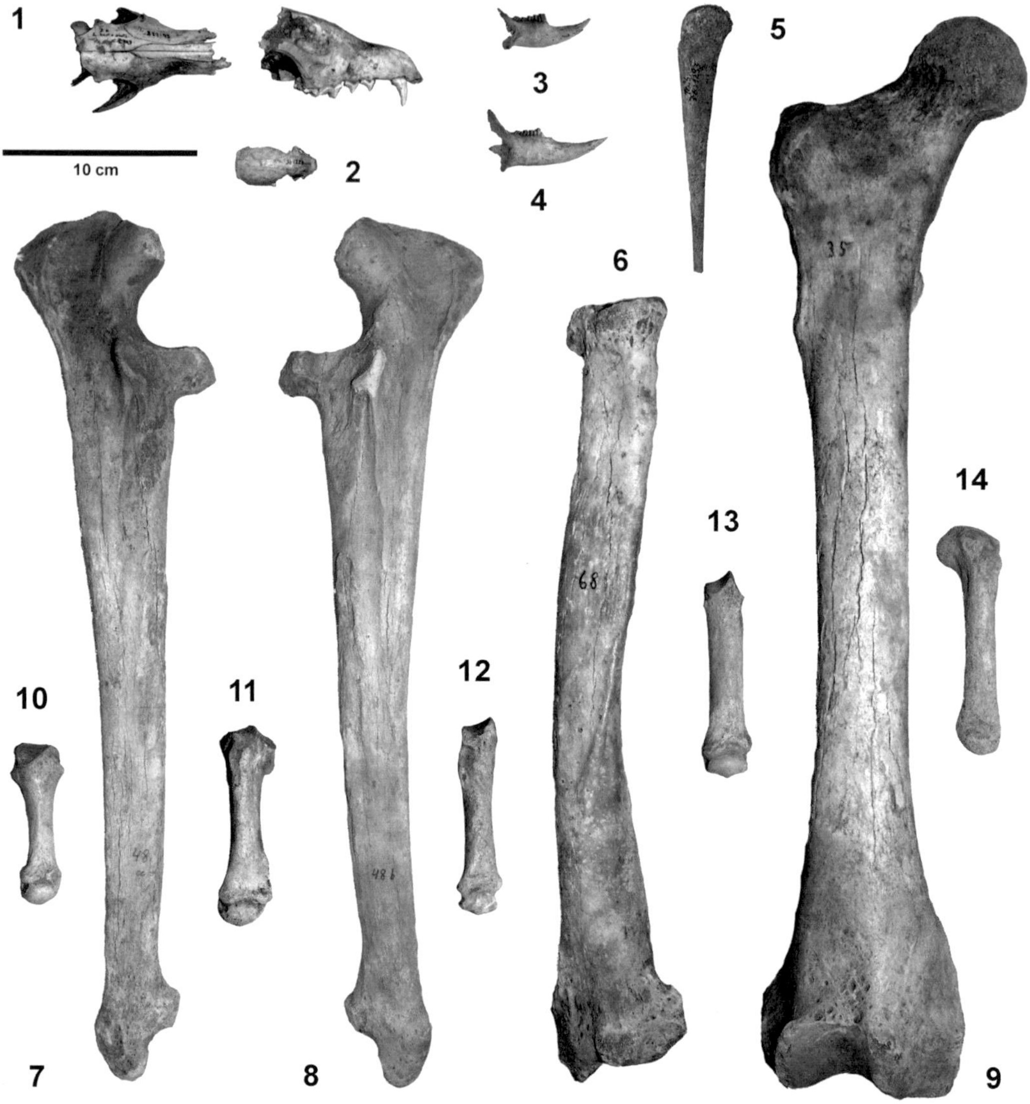

Figure 4. LGM and Post-LGM (= MIS2) fauna remains from the frost breccias cone area of the fist 5 meters of the former entrance and brown bear remains from the Lower Fluvial Passage of the Hermann's Cave. 1. Polar fox *Vulpes lagopus* anterior skull (LDA), dorsal and lateral. 2. Hermelin *Mustela* cf. *erminea* skull (LDA), dorsal. 3. Snow hare *Lepus timidus* mandibula (LDA), lateral. 4. European hare *Lepus europaeus* mandibula (LDA), lateral. 5. Prezwalski horse *Equus ferus przewalski* metapod (LDA), lateral. 6-14. Large grown up brown bear *Ursus arctos* subsp. leg remains of a single individual (composed from [6] with new adds).

Bode Valley fits to the typical den positions of polar foxes [19, 21]. In the loess, there must have been before the collapse of the sediment into the Baumann's

Cave system a burrow system, where they protected and bred their cubs [13]. The guilt bone accumulator was obviously *Gulo gulo*, who accumulated mainly reindeer remains (NISP = 520), whereas polar foxes mainly accumulated snow hare (*Lepus timidus*) and willow grouse (*Lagopus lagopus*) remains (Table **1**).

Further downstream, 20 meters above the today's Bode River, in the Fuchsloch Cavity there was an European eagle own den, which fauna consists mainly of birds (Table **1**), whereas a quantitative statistics still lacks [23]. Such *Bubo bubo* subsp. dens generally contain a variety of guilt bones of medium-sided mammals, birds and micromammals [24] similar as found in the Fuchsloch Cavity.

Table 1. Fauna of the LGM/Post-LGM pleniglacial along the Bode River Valley combined of the wolverine, polar fox, sable, hermelin and Eurasian eagle owl den bone assemblages (= predator guilt remains) in three Rübeland Caves (data composed and few modified from [13, 23]).

Species	From Den Site Type	Climate/Vegetation Zone
Herbivore Mammals		
Capra ibex	Wolverine/polar fox den	High alpine, rocky area
Equus ferus przewalskii	Wolverine/polar fox den	Tundra, open landscape, mountain valleys
Ovibos moschatus	Wolverine/polar fox den	Taiga, tundra
Rupicapa rupicapra	Wolverine/polar fox den	High alpine, rocky area
Carnivore Mammals		
Vulpes lagopus	Wolverine/polar fox den	Taiga, tundra, boreal forest
Gulo gulo	Wolverine den	Taiga, tundra, boreal forest
Mustela cf. *erminea*	Polar fox/least weasel/sable den	Boreal forest, rocky area
Mustela nivalis	Eurasian eagle owl den	Boreal forest, rocky area
Micromammals		
Apodemus sylvaticus	Eurasian eagle owl den	Ecological variable
Arvicola terrestris	Eurasian eagle owl den	Ecological variable
Barbastela barbastellus	Eurasian eagle owl den	Cave, ecological variable
Citellus superciliosus	Eurasian eagle owl den	Taiga, tundra, boreal forest
Clethrionomys sp.	Eurasian eagle owl den	Ecological variable
Cricetus cricetus	Polar fox, Eurasian eagle owl den	Tundra
Dicrostonyx henseli	Eurasian eagle owl den	Taiga, tundra, boreal forest
Eptesieus serotinus	Eurasian eagle owl den	Cave, ecological variable
Eptesieus nilssoni	Eurasian eagle owl den	Cave, ecological variable
Lemmus lemmus	Eurasian eagle owl den	Taiga, tundra, boreal forest
Lepus timidus	All den types	Taiga, tundra, boreal forest

(Table 1) contd.....

Species	From Den Site Type	Climate/Vegetation Zone
Microtus agrestis	Eurasian eagle owl den	Ecological variable
Microtus arvalis	Eurasian eagle owl den	Ecological variable
Microtus gregalis	Eurasian eagle owl den	Tundra
Microtus nivalis	Eurasian eagle owl den	Ecological variable
Microtus oeconomus	Eurasian eagle owl den	Ecological variable
Myotis sp.	Eurasian eagle owl den	Cave, ecological variable
Neomys fodiens	Eurasian eagle owl den	Ecological variable
Ochotona pusilla	Polar fox/sable, Eurasian eagle owl den	Tundra
Sicista subtilis-betulina	Eurasian eagle owl den	
Sorex araneus	Eurasian eagle owl den	Ecological variable
Talpa europaea	Eurasian eagle owl den	Ecological variable
Birds		
Aegolius funereus	Eurasian eagle owl den	Holarctic, boreal forest
Anas crecca	Eurasian eagle owl den	Ecological variable
Arthya fuligula	Eurasian eagle owl den	Ecological variable
Asio otus	Eurasian eagle owl den	Ecological variable
Bubo bubo subsp.	Eurasian eagle owl den	Mountains, and forests with rocky area, alpine
Circus cyaneus	Eurasian eagle owl den	Taiga, ecological variable
Columba palumbus	Eurasian eagle owl den	Ecological variable
Colymbus sp.	Eurasian eagle owl den	Arctic
Corvus corax	Polar fox, Eurasian eagle owl den	Ecological variable
Corvus menedula	Eurasian eagle owl den	Ecological variable
Lagopus lagopus	Polar fox, Eurasian eagle owl den	Tundra
Lagopus mutus	Eurasian eagle owl den	Alpine
Lyrurus tetrix	Polar fox, Eurasian eagle owl den	Palaearctic, alpine
Falco peregrinus	Eurasian eagle owl den	Ecological variable
Falco tinnunculus	Eurasian eagle owl den	Ecological variable
Mergus merganser	Eurasian eagle owl den	Holarctic
Picus canus	Eurasian eagle owl den	Boreal forest, ecological variable
Picoides tridactylus	Eurasian eagle owl den	Boreal forest, alpine
Perdix perdix	Eurasian eagle owl den	Ecological variable, mainly open landscape
Strix aluco	Eurasian eagle owl den	Boreal forest
Surnia ulula	Eurasian eagle owl den	Boreal forest

(Table 1) contd.....

Species	From Den Site Type	Climate/Vegetation Zone
Tetrao urogallus	Polar fox, Eurasian eagle owl den	Alpine, high mountains
Anures		
Rana temporaria	Eurasian eagle owl den	Ecological variable

A Hibernating Brown Bear

In the Hermann's Cave, from a single adult brown bear individual at least some of the postcranial bones are preserved, whereas most probably other material (especially cranium) disappeared during the first explorations of the cave after its discovery. The fore and hind leg bones figured herein (Fig. **4.6-4.14**) were collected from the surface in the anterior part of the Lower Fluvial Passage (Fig. **3**) [13]. Those represent remains of a large brown bear type, which is not yet well studied. It seems, those are from a bear that hibernated in the former entrance area, which finally died there, and was no further scavenged by wolves or smaller carnivores (absence of bite marks on longbones).

Landscape Change and Hermann's Cave Use

Drastic climatic changes after the Eemian Interglacial (MIS 5e) caused the initial change of the vegetation to a boreal forest in the Harz Mountain (Fig. **5A**) [13]. It was the time of the large cave bears *U. s. spelaeus* and smaller *U. s. eremus,* such as their Ice Age lion, spotted hyena and wolf predators, which latter specialized in such forests on cave bear feeding, whereas Neanderthals were present north of the Harz and to the Hermann's Cave opposite Baumann's Cave [13]. About 77.000 years ago (beginning of MIS 4) the Brocken Peak Ice Cap and valley glaciers must have started to grow within the increasing temperature drop (Fig. **5B**) [13]. Within the MIS 3, the Late Palaeolithic human hunter-gatherers arrived at Rübeland which hunted in the Hermann's Cave cave bears about 32.000 years ago (see Chapter Aurignaciens).

28.000 years ago the glaciation started again to increase by another temperature decline (Fig. **5C**) which finally was the motor for drastic changes, from a boreal forest to taiga forest, and artic-alpine conditions, which caused the extinction of the European mammoth and boreal forest fauna also in this region [13]. The Last Glacial Maximum reached its arctic conditions 19.000 years ago globally [15], when the Harz region landscape was "human-free" (Fig. **5C**), and an arctic/high alpine "reindeer fauna" replaced the former "cave bear faunas" of the Harz Mountains [13]. In the time until the end of the Ice Age (12.000 BP) arctic predators (wolverine, polar fox, mustelids, eagle owl) used partly the Hermann's Cave and other caves or sinkholes as dens in the Rübeland region along the Bode

Valley [13]. Humans returned not before 12.000 years ago with the last Epipalaeolithic reindeer hunters (Fig. **5D**). At this time, all larger caves portals, such as of the Hermann's Cave were already closed by frost breccias. Those hunter-gatherers found only small rock shelters on the Harz Mountain slopes for camp use (Fig. **5D**) [13].

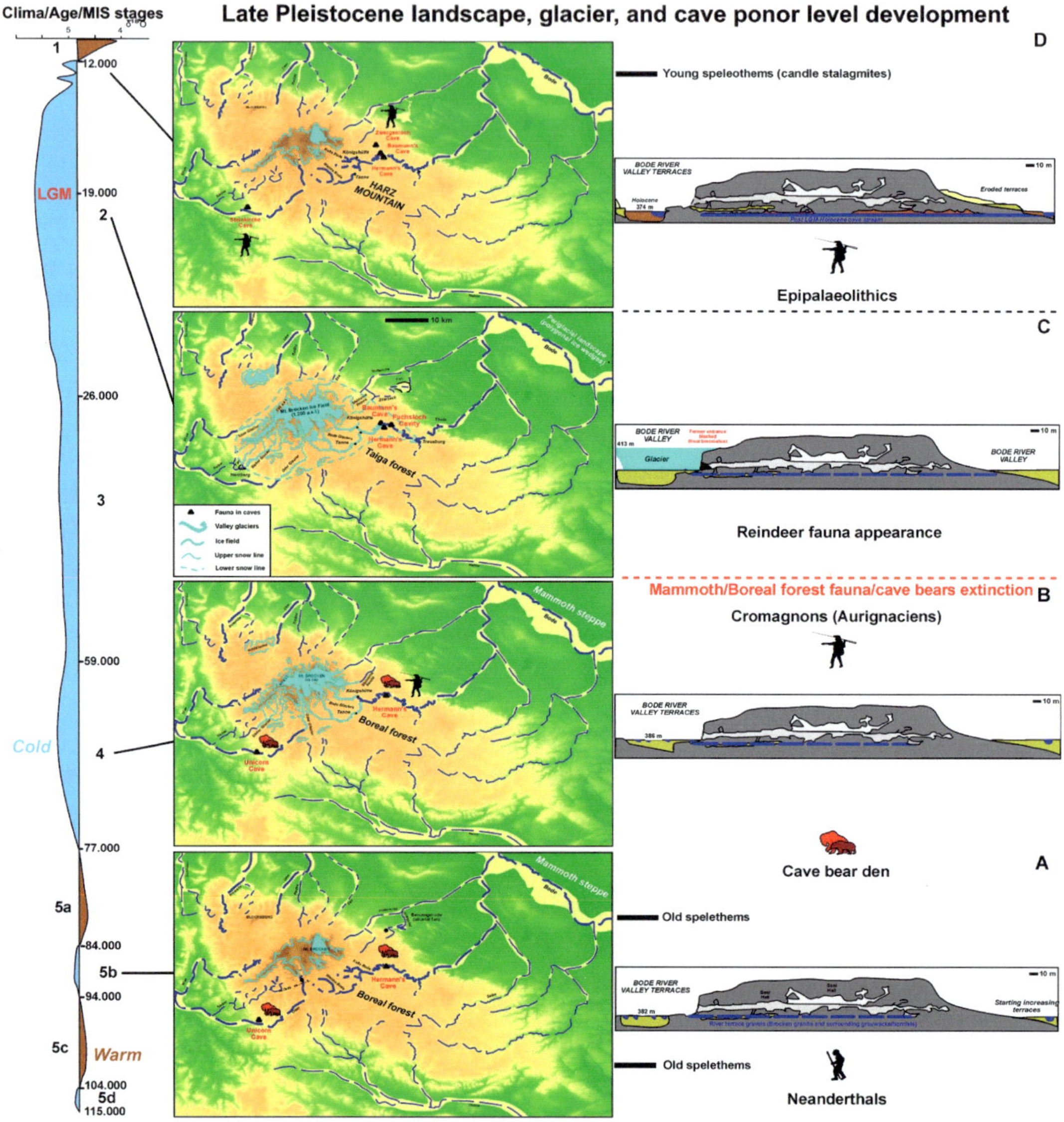

Figure 5. The Mt. Brocken Late Pleistocene glacier development model in four main stages - A-D - with the Bode Valley glacier and its impact onto the landscape (composed and modified from [13]).

REFERENCES

[1] Duphorn K. Ist der Oberharz im Pleistozän vergletschert gewesen? Eiszeit Gegenw 1968; 19: 164-74.

[2] Eissmann L. Das quartäre Eiszeitalter in Sachsen und Nordost-Thüringen. Altenburger Naturw Forsch 1977; 8: 1-98.

[3] Feldmann L. Das Quartär zwischen Harz und Allertal. Clausthaler Geowissenschaften 2002; 1: 1-49.

[4] Hövermann J. Die diluvialen Terrassen des Oberharzes und seines Vorlandes. Ein Beitrag zur Frage er Harzhebung. Petermanns Geogr Mitt 1950; 94: 121-30.

[5] Lembke H. Glazial, Periglazial und die eiszeitliche Schneegrenze im Harz. Geologie 1961; 10(4-5): 442-60.

[6] Vladi F. Quartärgeologische Untersuchungen zu den Terrassen der Sieber am Südwestrande des Harzes. Unpublished thesis, University Hamburg 1976; 1-109.

[7] Ricken W. Quartäre Klimaphasen und Subrosion als Faktoren der Bildung von Kies-Terrassen im südwestlichen Harzvorland. Eiszeit Gegenw 1982; pp. 109-36.

[8] Reinecke V, Lehmkuhl E. Untersuchungen zur Sedimentologie und geomorphologie der Terrassensequenz im nördlichen Harzvorland im Einzugsbereich der Ilse un Holtemme. Trierer Geogr Stud 2002; 25: 65-76.

[9] Weymann H-J, Feldmann L, Bombien H. Das Pleistozän des nördlichen Harzvorlands - eine Zusammenfassung. Eiszeit Gegenw 2005; 55: 43-63.

[10] Reinecke V. Untersuchungen zur jungpleistozänen Reliefentwicklung und Morphodynamik im nördlichen Harzvorland. Aachener Geogr Arb 2006; 43: 1-170.

[11] Bombien H. Geologisch-petrographische Untersuchungen zur quartären (früh-Saale-zeitlichen) Flußgeschichte im nördlichen Harzvorland. Mitt Geol Inst Univ Hannover 1987; 26: 1-131.

[12] Brüning H. Periglazial-Erscheinungen und Landschaftsgenese im Bereich des mittleren Elbetales bei Magdeburg. Göttinger Geogr Abh 1959; 23: 25-6.

[13] Diedrich C. Impact of the German Harz Mountain Weichselian ice-shield and valley glacier development onto Palaeolithics and megafauna disappearance. Quat Sci Rev 2013; 82: 167-98. [http://dx.doi.org/10.1016/j.quascirev.2013.08.013]

[14] Gibbard PL, Boreham S, Cohen KM, Moscariello A. Global chronostratigraphical correlation table for the last 27 million years Subcommission on Quaternary Stratigraphy. Cambridge, England: Department of Geography, University of Cambridge, Poster 2007.

[15] Clark PU, Dyke AS, Shakun JD, *et al.* The Last Glacial Maximum. Science 2009; 325(5941): 710-4. [http://dx.doi.org/10.1126/science.1172873] [PMID: 19661421]

[16] Skupin K, Speetzen E, Zandstra JG. Die Eiszeit in Nordwestdeutschland. Krefeld: GLA Nordrhein-Westfalen 1993; p. 143.

[17] Baranowski S, Henoch WE, Kucera RE. Glacier and Landform Features in the Columbia Icefield Area, Banff and Jasper National Parks, Alberta, Canada. Glaciology Division, Inland Waters Directirate, Environment Canada, 1978; p. 158.

[18] Reynolds SH. A monograph of the British Pleistocene Mammalia, II (III). The Canidae. Palaeont Soc Monogr 1909; pp. 1-28.

[19] Kuzmina IE, Sablin MV. Arctic foxes of the Late Pleistocene from the Upper Desna River. Trud Zool Inst RAN 1993; pp. 93-104.

[20] Pedersen A. Der Eisfuchs - *Alopex lagopus Linné*. Band 235. Neue Brehm-Bücherei, Wittenberg 2002; p. 44.

[21] Audet AM, Robbins CB, Larivière S. Alopex lagopus. Mamm Species 2002; 713: 1-10. [http://dx.doi.org/10.1644/1545-1410(2002)713<0001:AL>2.0.CO;2]

[22] Sheffield SR, King CM. Mustela nivalis. Mamm Species 1994; 454: 1-10. [http://dx.doi.org/10.2307/3504183]

[23] Arnold A, Böhme G, Fischer K-H, Heinrich W-D. Eine neue jungpleistozäne Wirbeltierfauna aus Rübeland (Harz). Wissensch Z Humb-Univ Berlin, math-naturw Reihe 1982; 31: 169-75.

[24] Andrews P. Owls, caves and fossils: predation, preservation and accumulation of small mammal bones in caves, with an analysis of the Pleistocene cave faunas from Westbury-sub-Mendip. Somerset, UK: University of Chicago Press 1990; p. 239.

CHAPTER 7

Late Palaeolithic Cave Bear Hunters

Abstract: The oldest record of Late Palaeolithic latest Aurignacien cave bear hunting and butchering is presented for the Hermann's Cave, a large Late Pleistocene cave bear den in the Harz Mountain (northern Germany), the most northern cave bear hunting (non-camp) site of Europe. Cromagnon man only occasionally hunted the largest cave bear species *Ursus spelaeus ingressus*, possibly also the small cave bear *Ursus spelaeus eremus*, with propulsion weapons in the larger cave systems. A 29.2010 ± 201 BP (latest Aurignacien/earliest Gravettien transition) old bone point (Aurignacien Mladeč type) and by use broken flint projectile were found both 90 meters deep in the upper cave bear bone/skeleton layers. Additionally, flint tools were discovered 40 meters deep, closer to the former entrance: two blades (one broken, both with lateral use) and a broken blade scraper, such as a flake scraper. The used damaged flint artifacts represent a butchering tool kit that was used for dismembering a cave bear carcass at the kill site. Two cave bear bones with deeper V-profile shaped cut marks were found, which correlate to damage use from one or two blades. Sickle-like parallel cuts are found on a femur shaft of an adult bear (*U. s. ingressus*). Straight cuts are on a scapula of an older neonate cave bear. Whereas in lowlands and river valleys of middle high mountain slopes in summer to the mountains migrating reindeer (and also horses) were the main big game target, in winter times in boreal forest mountain regions cave bears were hunted by Late Pleistocene humans, in competition with steppe lions, leopards, Ice Age spotted hyenas and wolves.

Keywords: Cave bear hunt, Late Aurignacien, Hunting 40-90m deep in cave bear den, Damaged propulsion projectiles, Butchering flint tool kit, Cut marks, Large cave bear *U. ingressus*, Most northern cave bear hunt record of Europe.

AURIGNACIENS IN EUROPE AND GERMANY – CAVE SITES

The first modern European humans migrated from northern Africa to central Europe with the Cromagnon humans around 36.000 BP [1] with most northern camp sites in England. About 50 secure absolute and by artifact typology dated sites are known [2 - 5]. Aurignaciens were spread all over Europe, from the Iberian Peninsula to the Caucasus, and to northern Africa [2, 6, 8 - 16]. The most northern cave site in Europe is the herein described German Herman's Cave. It is situated below the 1.142 meters a.s.l. high Brocken peak of the Harz Mountain. This is added to the palaeodemographic map of the Aurignacien period, especially

Cajus G. Diedrich

to those sites, which have signs of cave bear hunting concentrating in Central European boreal forests (Fig. **1**).

Several other German sites of Aurignaciens are described mainly from the rock shelter entrances in southern Germany, especially the Ach Valley [4, 7]. Some German single artefact remains are controversial discussed, whereas in Saxony-Anhalt in northern Germany, only two localities are listed with Breitenbach open air [17], and the Hermanns's Cave.

Figure 1. Most northern distribution limitation of the Late Palaeolithic Cromagnon modern humans (upper black stippled line) in Europe with the Harz Mountain as most northern site in Europe. Cave bear den sites with Late Aurignacien/possibly Early Gravettien).

HERMANN'S CAVE EXCAVATIONS AND MATERIAL

The history of the Mr. Blasius excavations [18, 19] in the Hermann's Cave is compiled in more [20, 21] detail in the introduction chapter. The archaeological material comes from stratigraphic context – the upper 40 cm of the upper cave bear "bonebed" layer. Mrs. Steiner excavated in 1961 with figuring the sections where a few flint artifacts were found in the Bear Cave Hall [22]. Another flint blade was excavated together with a bone point in the Saal Hall [23] (Fig. **2**). Mr. Völker did some square excavations 1985 [24] directly besides the excavations of Steiner in the Saal Hall. Their sections with new redrawing, stratigraphical/ sedimentological new dating was compiled for the Hermann's Cave with its relevant data [20] (Fig. **3**).

Institutional abbreviations: The macrofauna (see Chapter Cave Bears, Table **1**) and archaeological remains (Fig. **4**) from the Hermann's Cave were studied in the archaeological and palaeontological collections of the Landesmuseum Denkmalpflege und Archäologie Halle/Saale (= **LDA**), Landesmuseum für Naturkunde Braunschweig (= **LMNB**), Naturkundemuseum der Humboldt-Universität Berlin (= **MB**), and Rübeland Caves (= **RC**), there, partly *in-situ*.

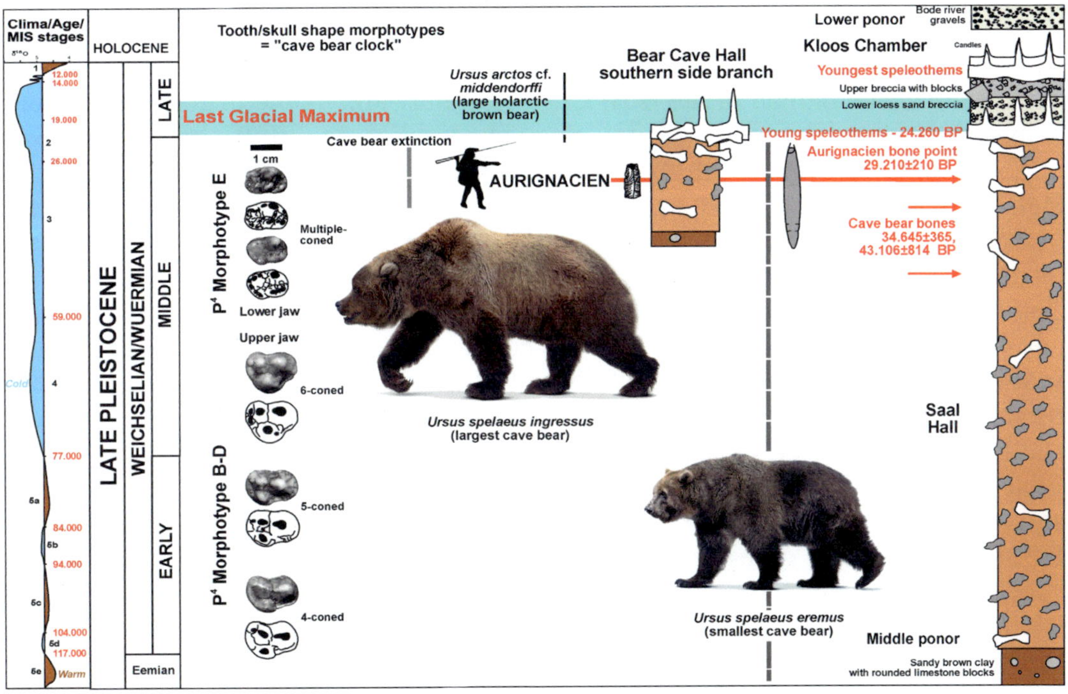

Figure 2. Generalized stratigraphy of the middle cave bear den ponor level and absolute dates on speleothems and of the Hermann's Cave with Late Palaeolithic (Late Aurignacien) remains. Cave bear species/subspecies and indirect dating by P4 morphology = "cave bear clock" (composed and modified from [20, 21]).

DATING OF ARCHAEOLOGICAL FINDS

Sedimentology, glacial cave genesis (Elsterian/Cromerian Glacial, Holsteinian

Interglacial, Saalian Glacial, LGM and to Holocene stages), cave bear subspecies evolution (= cave bear clock, cf. Fig. **3**) such as general bone taphonomy and species NISP analysis were performed [20, 21] and are presented in detail in different chapters in this book. Outside the caves, signs of the regional glaciations of the Late Pleistocene time frame (LGM stage around 19 ka) [20, 21] were used to understand the ponor cave stages. This interdisciplinary work outside cave (LGM-glacier geomorphology – valley glacier reconstructions) and inside to the glacier groundwater level related ponor elevations (Fig. **2**) at three main larger caves of the Harz Mountain Range (Baumann's, Hermann's and Unicorn Cave [20]), was important to understand the stratigraphic and taphonomic context of the Late Palaeolithic Aurignacien artifacts of the Hermann's Cave (Fig. **4**).

All artifacts have been excavated in the upper layers of the cave bear "bonebed/skeleton layer" [22] (Fig. **3**). These are non-stratified autochthonous clays, including a few rounded, weathered limestone blocks and abundant articulated, partly articulated or scattered skeletons (bones) of the two different sized cave bears, the smaller cave bears *Ursus spelaeus eremus* and largest form *Ursus spelaeus ingressus* [21]. In the Saal Hall, these layers are two meters thick, whereas in a side branch of the Bear Cave Hall they are visible in an open section, being represented by only 40 cm cave clays (Fig. **3**). Wolf, hyena and few imported prey remains were found only in the anterior cave part, the Bear Cave Hall [21] (Fig. **2**). These cave sediments were deposited within the Late Pleistocene [20]. The cave was dry and perfect for the hibernation for cave bears (= den), whereas the active ponor stream was 8-10 meters below [21] (cf. Fig. **3**). The ponor stream left different fluvial deposits, consisting of sediments: up to 5 cm large Bode River terrace gravels consisting of Palaeozoic volcanic and sedimentary rock types, which contain no bones. The artifacts are indirectly dated by the cave bear clock and cave bear species/subspecies using the P4-tooth and skull shape types [21]. Especially the presence of *U. s. ingressus* fits well to the final pre-LGM Late Pleistocene [21] (Fig. **3**). A cave bear bone sample on a single C14-dated bone (Bear Cave Hall) from the Hermann's Cave of 30.761±315 BP [25] (Fig. **3**) dates exactly into the Late Aurignacien [3, 26]. The "bonebeds/skeletons" and artifacts were finally covered starting around 25.000 BP by a younger speleothem generation that was absolute dated on the typical late speleothem genesis candle stalagmites between 24.260 ± 568 to 23.130 ± 711 BP [27] (Fig. **3**). Since the Gravettien, when sipping water made the cave non-attractive, and when cave bears became extinct [21], the cave became no longer accessible for Palaeolithic humans when its entrance collapsed and became blocked by frost breccias. Additionally it was a dangerous time in the cave, because of large ceiling collapses which date to the pre-LGM/LGM and early post-LGM times, these fallen blocks and stones finally covered and damaged larger parts of the bonebed areas and also skulls/bones [20, 21].

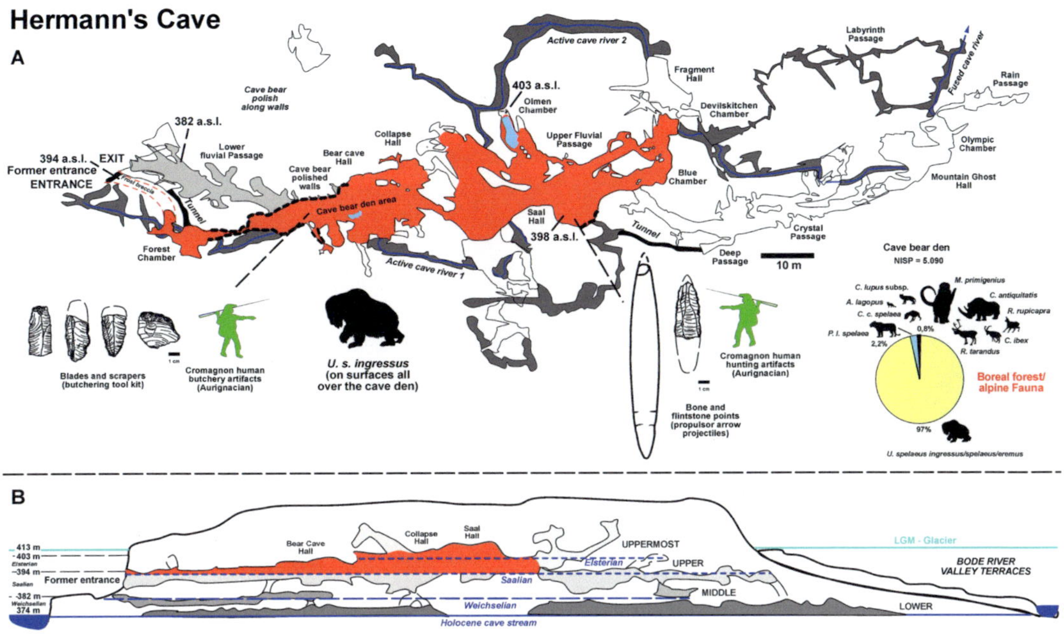

Figure 3. Hermann's Cave archaeology map. A. The large cave bear *Ursus spelaeus ingressus* cave bear den area (red) in birds-eyes view with distribution of six Late Aurignacien artefacts which have been found 40 meters deep in the Bear Cave Hall (butcher tool kit, and bones with cut marks: cf. Figs. **4**, **5**) and 90 meters deep in the Saal Hall (bone and stone projectiles: cf. Fig. **4**). B. During the Late Pleistocene, the active ponor stream was 8-10 meters below the dry cave bear den area (red) (composed and modified from [20, 21]).

AURIGNACIEN ARTIFACT DISTRIBUTION WITHIN THE HERMANN'S CAVE

The few Late Aurignacien artifacts are for the first time compiled all herein: one bone point, and five Nordic Flint-stone tools. Those different tool types were separately excavated in former times [22, 23, 28] and where discovered in two main areas (Fig. **3**). Both projectiles, bone and flint points (Figs. **4A-B**), were found in the Saal Hall, 90 meters distance from the entrance deeper in the cave; B. Four butchering tools with scrapers and blades (Figs. **4C-F**) were found 40 meters away and closer to the former entrance of the Bear Hall (Fig. **3**). The artifacts are based on the sedimentological/cave bear taphonomy results in-situ finds. Indeed, it cannot be fully excluded at this time, that cave bears imported the two propulsion projectiles, after being shot (projectiles sticking in their flesh) into the Saal Hall, where they died, finally. The butchery tool kit from the Bear Hall must be in full autochthonous context.

LATE AURIGNACIEN TOOL KIT FROM THE HERMANN'S CAVE

Two artifacts (Figs. **4E and D**) represent unpublished specimens. The compiled material allow a new or updated interpretation of the "archaeological site" and its context to the "cave bear den site".

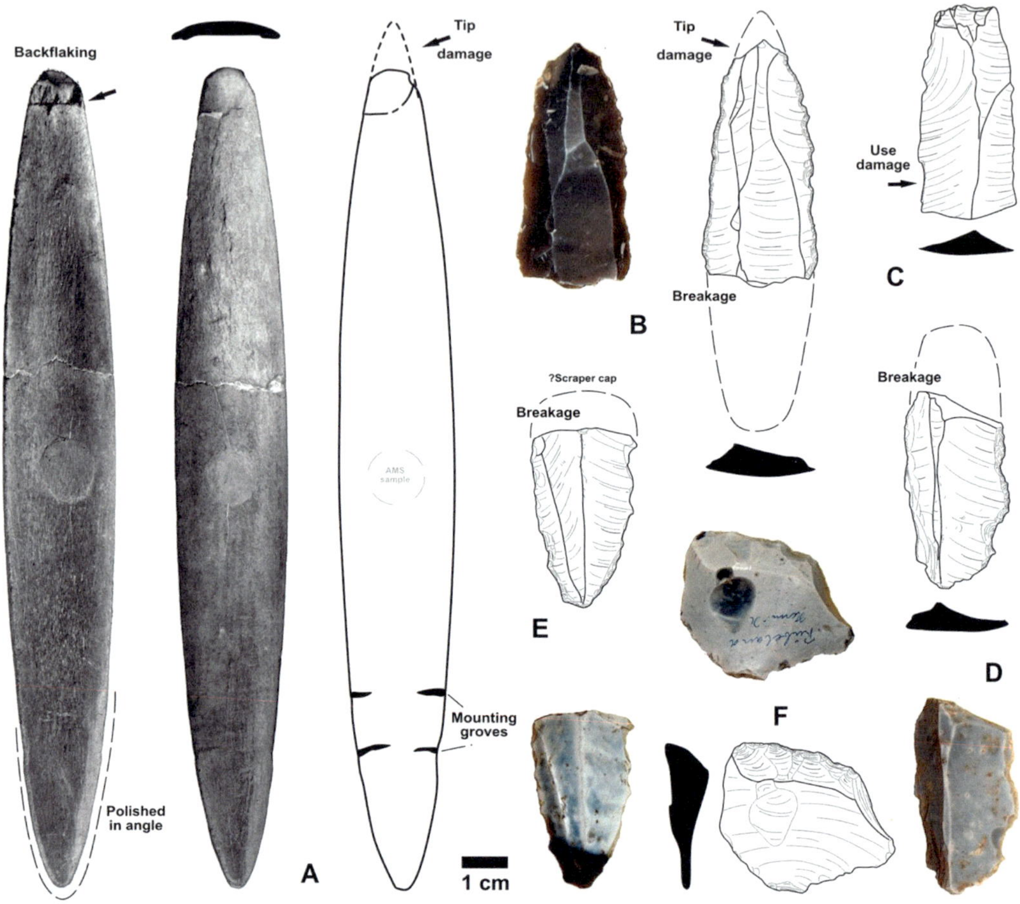

Figure 4. Late Palaeolithic Late Aurignacien cave bear hunting tool kit – bone point, and flint point, blades and scraper artefacts (see their finding positions in Fig. **3**) from the Hermann's Cave at Rübeland, Harz Mountain, northern Germany. A. Lautscher/Mladeč bone point from the Saal Hall (LDA no. HK 63:145, photos modified from LDA). B. Blade with Aurignacien retouch (= Aurignacien point) from the Saal Hall (LDA no. HK 56:286). C. Flint-projectile, blade with typical Aurignacien retouches (original lost, redrawn after [28]) from the Bear Cave Hall. D. Flint-blade from the Bear Cave Hall (LDA no. HK 84:938,1). E. Flint endscraper on blade with Aurignacien retouch from the Bear Cave Hall (LDA no. HK 84:938,2). F. Flint-scraper (LDA no. HK 63:144) from the Bear Hall.

The Lautscher/Mladeč bone point from the Hermann's Cave is 186 mm long with a maximal width of 27 mm, a thickness of only 6 mm and 32.2 g in weight (Fig. **4A**). The larger AMS sample in the center of the artifact was completed. The bone point tool has been dated AMS 29.210 ± 210 BP [29], whereas other absolute dated bone points were compared to absolute dated culture layers of similar aged

Aurignacien sites of Europe (*e.g* [4, 26].). A backflaking on the opposite side of the former tip demonstrates damage by use. The tip damage, and especially basal angular polishing and basal double and not deep grooves (Fig. **4A**) indicate its use as figured upside. The origin of the raw material must be identified as, most probably, reindeer antler. The "compacta" is about 2 mm thick, which is more typical of reindeer than other cervid antlers (*e.g.* red deer). Furthermore, the internal spongiosa structure has typical elongated antler structures, which allow the exclusion of it being made of longbones. Compared to other Aurignacien bone points of Europe [7, 15, 30 - 33], these are made, *e.g.* at Isturitz and Tuto de Camalhot (F) or Mammoth Cave (Pl) sites, of reindeer antler, mammoth tusk ivory and ribs, and less of megafauna longbones [34]. Reindeer antlers with smooth surfaces are well known to have been used in all Late Palaeolithic periods for propulsion bone points [32, 33, 35, 36]. Similar Lautscher/Mladeč bone point fragments have been dated in similar Late Aurignacien age and were reported further south from the Greek Klisoura Cave bear den/archaeological site that was used as a camp site by Cromagnons in its entrance area [26]. Other bone point fragments have been figured *e.g.* from the type site Mladeč Cave [15], three caves in southern Germany such as Vogelherd, Hohlestein Stadel and Hohle Fels Caves [4, 37] or Polish Mammoth Cave demonstrating a more south-eastern European-wide distribution of these hunter groups [30] (cf. Fig. **1**).

All <u>five flint tools</u> from the Hermann's Cave are made of Nordic grey-black flint and have a more or less white-bluish patina (Fig. **4**) as a result of surface interaction with carbonates in the cave sediment (Devonian limestone). The raw material for the stone tools must have been collected north of the Harz Mountain in the Lowlands (flint from Saalian river terrace deposits, cf [38].). Earlier, in the Middle Palaeolithic, Neanderthals used the same Saalian transported flint types from the Bode River Gravels downstream, not far from Westeregeln [39, 40]. The Aurignacien flint artifacts were imported as tools (or raw material with an unknown production place) nearby Bode River upstream from the cave, because the local terraces in the Harz Mountain Range do not contain any flint material.

The <u>blades and blade scraper</u> of the Hermann's Cave are all regular and large types of typical 20-30 mm wide Late Palaeolithic blades (not lamellas). The blade forms of the Hermann's Cave fit into the general morphology of regular and large Aurignacien (Late Palaeolithic) blade techniques, such as the types and places of the rework [4, 8, 13, 31, 37, 41, 42]. Three of the blades (Figs. **4C-E**) have no lateral retouch but have small fractures on the cutting edges, indicating use on "hard material" (most probably bone). One of these has the bulbous area (typical blade), the other one might reflect a broken blade with the bulbous missing. Another blade (Fig. **4B**) has the typical Aurignacien lateral retouch which identifies it best as a typical "Aurigancien blade". The other shorter and more triangular blade has no retouch, but is also broken (Fig. **4E**). This one seems to represent a fragmented end scraper on blade with Aurignacien retouch. Another larger flake scraper has three flake negatives on the dorsal surface (Fig. **4F**). The scraper reduction was made in the former bulbous area. The blade tool fragments can be compared, best fitting to similar fragmented tools from the French Hyena Cave Brassempouy, Arcy-sur-Cure Cave or Caminade West (F) [42, 43], the Mammoth Cave (Pl) [30], or the Hohlestein Stadel Cave (D) [31]. The "Aurignacien blades with lateral retouche" [4] and herein Fig. **5B**) seem to represent propulsion spear projectiles. Compared to the younger Gravettien flint projectiles, these are similarly large and have similar retouch, or damaged tips. The function as area projectile point explains why they are often being preserved only partially (mainly half). This would explain both damages on the projectiles of the Hermann's Cave, tip damage and breakage (Fig. **5B**).

All tools (two projectiles, two blades and two scrapers) are in keeping with the technology and industry of Aurignacien stone tools, from the Czech Republic Early and Late Aurignacien Mladeč Cave [15], Polish Late Aurignacien Mammoth Cave and Deszczowa Cave [30], German Geißenklösterle, Hohlestein Stadel (Early Aurignacien), Vogelherd Cave (Early-Late Aurignacien) and Hohle Fels Caves [4, 31] and the Greek Klisoura Cave [26], which were in all cases also overlapping entrance camp sites and cave bear dens.

CUT MARKS ON THE CAVE BEAR BONES

The bones from the large bone dump residue of the historical excavations by Kloos and Müller in 1889 [44] at Bear Cave Hall were checked for carnivore bite marks and damages. Two cave bear bones from the Bear cave Hall, where the butchering tool kit was also found (Fig. **2**) were discovered, which have sinuous elongated scratches and drop-shaped scratches (Fig. **5**). All marks were compared to exclude possible "modern damage" origin and ancient "human *versus* carnivore" origin (carnivore damage criteria *e.g.* in [45, 46]). This is possible by the cross-section form V-shapes (cut marks) *versus* U-shapes (bite marks) sensu [47, 48], and the surface outline, which form was distinguished (curved and straight = cut marks) from sinuous or drop-shaped elongated bite marks (= carnivore, *e.g* [45, 48].). The different shaped cut marks (Fig. **5**) were studied under a binocular (U- *versus* V-shape).

The cut marks found at the femur of the Hermann's Cave have straight or convex surface forms (Fig. **5**). About eight stepped convex scratches, with distances of about 5-12 mm on the femur of an adult animal (smaller sized *U. s. ingressus* female), are 5-25 mm long (Fig. **5A**). Between, there are two sinuous-like and U-shaped short carnivore bite scratches, which also have very different orientations

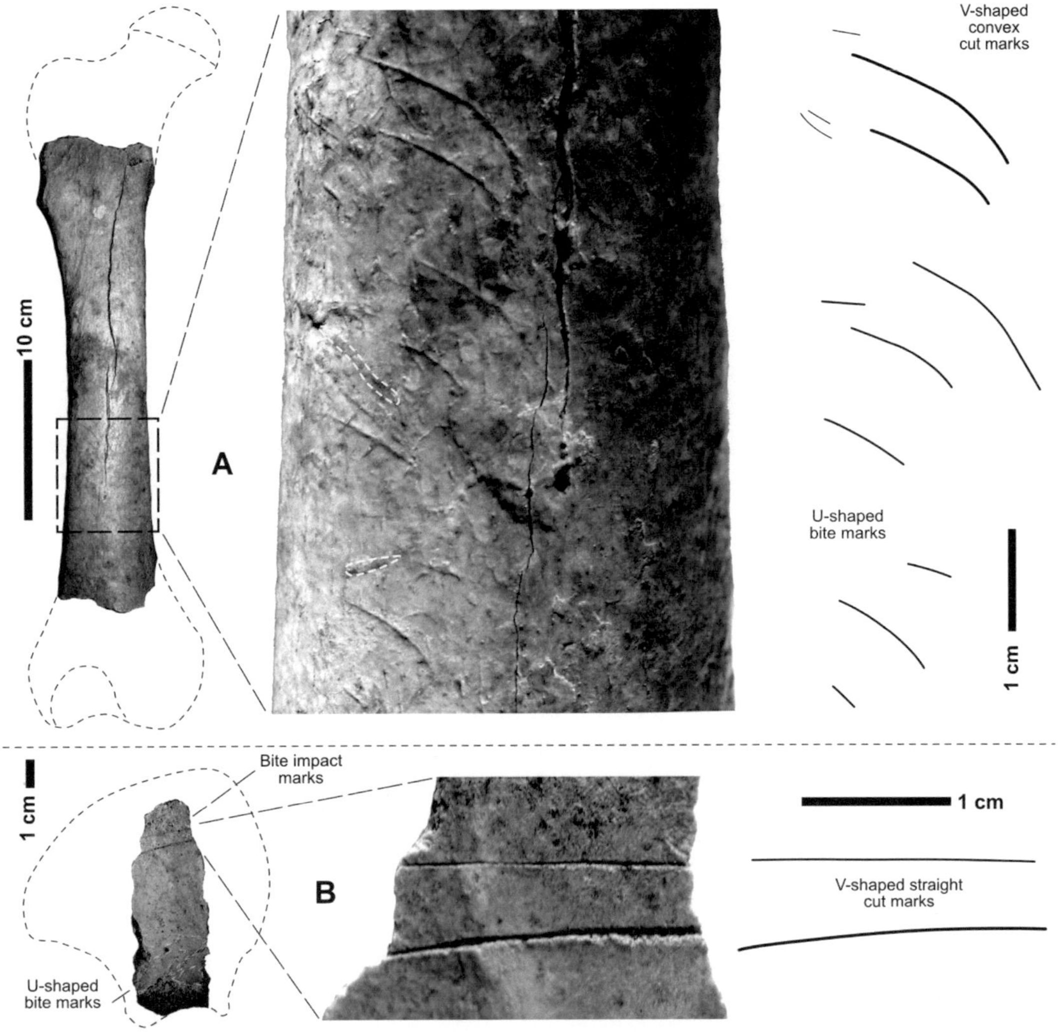

Figure 5. Bones with Late Palaeolithic cut marks from the Bear Hall of the Hermann's Cave, Harz Mountain, Saxony-Anhalt, northern Germany. A. Right femur shaft (historically damaged joints) of an adult animal of the large cave bear species *U. s. ingressus* (most probably female) with curved parallel scratches and few smaller carnivore (possibly fox/badger) bite marks (RC no. HC BF-1). B. Scapula of an older neonate with two parallel straight cut marks and carnivore bite mark/damage (RC no. HC BF-2).

compared to the cut marks. All cut marks are V-shaped in their profile, such as the nearly parallel and straight marks found on the inner side of a shoulder blade of a neonate cave bear cub (Fig. **5B**). These marks are even deeper, which might be mainly explained that the cub bone was less calcified, different as to the femur shaft compacta of an adult cave bear. Similar V-shaped and curved cut marks were compared to several ones (interpreted as "ornamentation") which were found on a "megamammal" bone (probably mammoth/rhinoceros rib fragment) at the Aurignacien (Gravettien) open air site Breitenbach (cf. photo in [49]), whereas there also a typical similar to the Hermann's Cave Aurignacien bone point fragment (lower half) was found [50].

Following arguments of cut mark positions in general [47, 48], the anterior middle shaft position of the cut marks on the cave bear femur from the Hermann's Cave could result from "filleting", which might have been caused either by a blade (straight forms), or also by a scraper (convex forms, cf. experiments in [47, 48]).

The straight cut marks on the neonate cave bear scapula were caused by butchering and body cavity opening of a pregnant female (probably when the hunters removed the intestines/inner organs with their flintstone blades). Both bones would fit to a winter game hunting scenario and the easy kill of a pregnant female during hibernation. The proportions of the femur fit better to a smaller female *U. s. ingresssus*. After butchering, a few bite scratches from a smaller carnivore (most probably fox/badger) (Fig. **5A**) demonstrate the recycling of the butcher remains.

From several caves of Europe "cave bear hunt" reports of Late Paleolithic's are based only on cut marks on cave bear bones or drilled cave bear teeth [16, 51, 52]. In southern Germany a drilled cave bear canine was found within the Aurignacien layers [7], whereas other Aurignacien/Gravettien drilled incisive teeth, canines with cut-lines, or canines as retoucher and two sawed fibula bones were figured from the French Pyrenean Gargas Cave [52]. Cave bear teeth were used as pendants in Aurignacien jewelry [4, 15, 53]. Drilled or cut cave bear teeth from the German Hohlestein Stadel Cave [31] and Polish Deszczowa Cave [30] seem to have a special symbolic character. Possibly these were carried only by the cave bear hunters (or shamans), which took them as "trophies" of their prey.

Cave bear shamanic ochre colorations have been reported on cave bear bones and skulls all over Europe [16], and possibly by Late Palaeolithic shamanic positioning of skulls (*e.g.* Peters Cave – [54], Chauvet Cave – [55]), does not support their hunting by humans. Cave art (Aurignacien/Gravettien) in several SW-European caves include figures of cave bears (or even brown bears) interpreted as "hunted cave bears with spears" that are also controversially discussed, whereas those spears might represent possibly propulsion spears [16, 56]. That the cave bear played a role in Late Palaeolithic shamanism, without a doubt, is further documented at the Aurignacien Hohle Fels Cave site with the "bear-human" figure (possibly a shaman) in a "cave bear or with newer interpretations lion fur" [57]. Furthermore, a cave bear ivory figure was found in the Aurignacien layers at the southern German Geißenklösterle Cave [31], such as at the Gravettien/Pavlovian Czech Republic Dolni Vestonice I site [58] demonstrating the relevance of cave bears for the early modern European humans. Cave bear bones with cut marks from a Gravettien layer (Epigravettien, 26.200±300 BP [30]) in southern Germany [30] are similar to French or Belgian cave beasssr bones found at 16 Late Palaeolithic sites (cf [59].). Similar cut marks (including a cut male cave bear canine) were also described from the Late Aurignacien site of Deszczowa Cave, Poland [30]. Bones with cut marks could have been collected in large amounts from the surface bonebeds in caves by Late

Palaeolithics and could have been simply used for bone tool or jewelry purposes, or shamanistic bone depots (from the cave floor reused cave bear bones). Therefore, this is not the best evidence of "cave bear hunting".

AURIGNACIEN CAVE BEAR HUNTING SPECIALIZATION IN MOUNTAIN REGIONS

Although Aurignacien megafauna prey of an increasingly colder climate within the Last glacial period varied, reindeer were the most important game followed by horses. This has only been studied in south-western France [60 - 62], where the climate was warmer, as in the herein studied Harz Mountain region that was much closer to the Scandinavian Inland glacier. The Harz Brocken Peak became covered by an Ice Cap starting in the Gravettian/Solutréen, therefore there was no Late Palaeolithic human occupation in this region [20]. The big game hunting specialization is established for lowland and several camp sites along river valleys reaching to mountain slopes of middle high mountain regions. There, reindeer herds migrated to the mountains in the summer, and again back to the lowlands in the autumn [63]. Therefore, other prey specializations in different landscapes - boreal forest of middle high mountain regions - must be expected, where some big game was rare or absent, such as the mammoth [20, 21, 29]. Not only humans had to adapt in such "non-mammoth steppe landscapes" on different game, similar as their competitors: steppe lions, leopards, Ice Age spotted hyenas and Ice Age wolves. Those top predators are recently well studied (see also chapter lions) and known to have specialized in boreal forest zones to hunt or scavenge cave bears deep in caves [20, 21, 64 - 66], such as described in this book for the Hermann`s Cave. Because of the predation stress of human and carnivore hunters, cave bears hibernated in most cases as deep as possible in cave systems with best example found in the Romanian Urşilor Cave [67] or the herein discussed Hermann's Cave (see chapters cave bears, lions).

In this contribution, the oldest records for cave bear hunts by modern weapon technology, propulsion spears of early Late Palaeolithic Aurignaciens is presented. After the controversial discussion about the Aurignacien/Gravettian boundary around 30-29 ka [3], the Hermann's Cave is slightly younger in age or at the final end of the Aurignacien. Further convincing evidence for cave bear hunts comes finally much more from the younger Gravettien, in which a dorsal vertebra of a cave bear from the southern German Hohle Fels Cave Gravettien camp site at a cave entrance contains a flint projectile fragment [30, 68, 69]. This bone was dated by C14 27 830 ± 140/150 BP [69] and is from the large and last European cave bear species *Ursus spelaeus ingressus* [53], similar as in the Hermann's Cave.

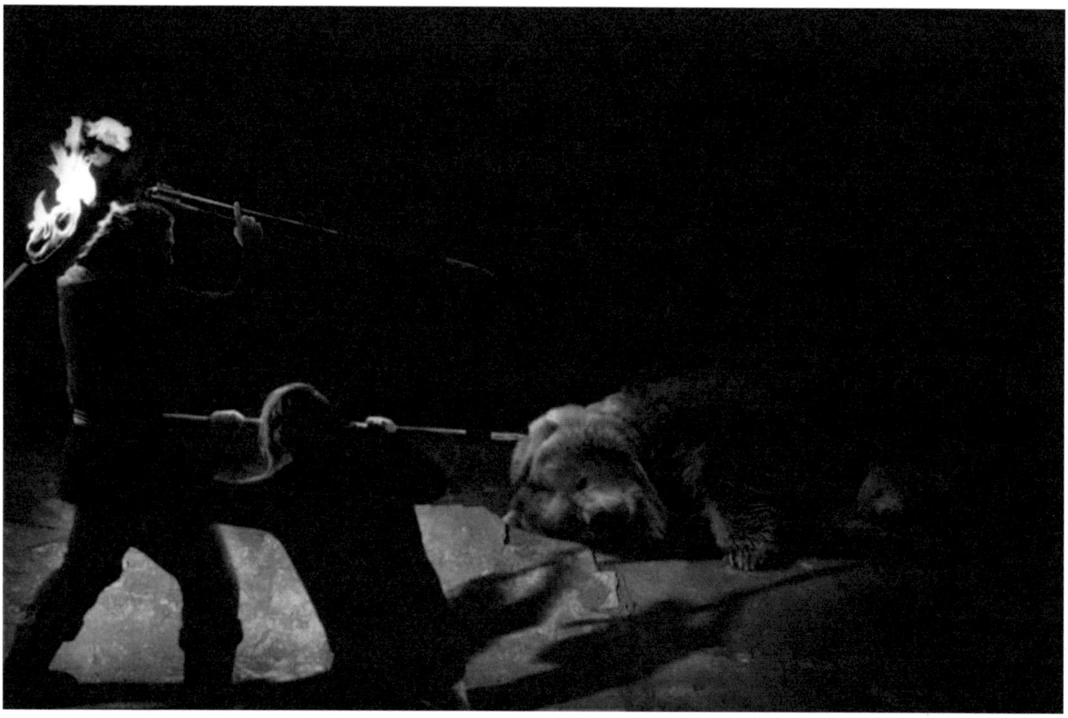

Figure 6. Possible hunting scenario. A. Late Palaeolithic Cromagnon modern humans (Late Aurignacien) hunt in a small group for cave bears deep in caves about 29.210 years ago (dated by the bone point) during winter times, when cave bears hibernated most vulnerable (Illustration R. Uchytel).

Hunting Deeper in Cave Bear Den Sites - Possible Hunting Strategy Scenario

Another possibility besides direct proof (projectile pieces in bones) is the artifact scattering mapping in cave bear dens, and herein it is quite unique in the Hermann's Cave for the European record. The Hermann's Cave represents now with this book presentation one of the best interdisciplinary studied German cave bear dens, with most convincing evidence of Late Aurignacien cave bear hunting and butchering at the killing site and deeper in caves (Fig. **2**).

First the cave was discussed to have been used as "probable inhabitation" [17] ignoring the autochthonous origin of the cave bear bonebeds/artifacts, even deep in the cave (cave bear den area, Fig. **2**). Controversially, in former times proposed all Aurignacien artifacts were believed to have been "washed" into the cave [17]. Their autochthonous presence was proven with the detailed interdisciplinary studies and especially the sedimentology (see chapter II). An "inhabitation" of the Hermann's Cave in its deeper parts was obviously not present (few artifacts, no work debris of stone tool production, no fire places, *etc.*). Possibly, there is indeed additionally a "camp site" at the former entrance before it was covered by frost breccias around the LGM time (cf. Fig. **2**). The 8-10 meters of this unexplored first cave part is still uncovered and might contain a speculative "camp site" situation. It is highly questionable, as the cave was mainly used as a cave bear den up to its former entrance. Also it was possibly used as sporadic hyena and wolf dens [21].

Hunting cave bears in this period by humans was mentioned, based singularly on the presence of the Lautscher/Mladeč point found deeper in the cave [29]. The central European cave sites Mladeč Cave, Mammoth Cave, Hohlenstein Stadel Cave, Vogelherd Cave, Hohle Fels Cave [4, 15, 30, 31] (Fig. **1**) have similar tool kits and have also cave bear bones that expose cut marks. These are not large cave bear dens, but those are overlapping camp sites/animal dens in cave entrances with "cultural layers". Artifacts have been found there in the entrance areas only and not deep in such caves, which is different in the Hermann's Cave. Aurignacien camp sites in mountain regions are, in all cases, close to, but not at the cave entrances of the cave bear dens; *e.g.* the Hohlenstein Stadel Cave Aurignacien camp site was close to the Hohlestein Bear Cave, a larger cave bear den [4].

Aurignaciens most probably hunted in small groups in caves, which is a different strategy than those proposed for large mammoth steppe hunts with larger groups in lowlands or river valleys (*e.g* [36].). Here in the cave darkness, a most probable speculative scenario is presented in an night-google vision illustration (Fig. **6**) where a small group (around three hunters) might have made up a hunting team; at least one for lighting, one for spear throwing, and possibly one for carrying the weapons and butchering equipment.

REFERENCES

[1] Henry-Gambier D. Les fossiles de Cromagnon (Les Eyzies-de-Tayac, Dordogne): nouvelles données sur leur position chronologique et leur attribution culturelle. Paléo 2002; 14: 201-4.

[2] Kozlowski JK. The formation of the Aurignacien in Europe. J Archaeol Res 2000; 56: 513-34.

[3] Dinnis R. The archaeology of Britain's first modern humans. Antiquity 2012; 86(333): 627-41. [http://dx.doi.org/10.1017/S0003598X00047815]

[4] Conrad NJ, Bolus M. The Swabian Aurignacien and its place in European Prehistory. Trab Arqueolog 2002; 24: 211-39.

[5] Conard NJ, Bolus M. Radiocarbon dating the appearance of modern humans and timing of cultural innovations in Europe: new results and new challenges. J Hum Evol 2003; 44(3): 331-71. [http://dx.doi.org/10.1016/S0047-2484(02)00202-6] [PMID: 12657520]

[6] Perpère M. Grands gisements Aurignaciens de Charente. Anthropologie 1975; 75(2): 243-60.

[7] Hahn J. Aurignacien. Das ältere Jungpaläolithikum in Mittel- und Osteuropa. Fundamenta 1977; A9: Köln und Wien.

[8] Djinjian F. Recherches sur l'Aurignacien du Périgord à partir des données nouvelles de la Ferrassie. Anthropologie 1988; 90(1): 89-106.

[9] Djindjian F. Les origines du peuplement aurignacien en Europe. In: Bánesz L, Kozlowski JK, Eds. Actes du XIIe Congrès International des Sciences Préhistoriques et Protohistoriques. Bratislava. 1991; pp. 136-54.

[10] Gambassini P. L'Aurignacien de l'Italie du sud. In: Bánesz L, Kozlowski JK, Eds. Actes du XIIe Congrès International des Sciences Préhistoriques et Protohistoriques. Bratislava. 1991; pp. 203-6.

[11] Cabrera Valdés V, Bernaldo de Quirós F. L'Aurignacien de la région cantabrique espagnole. In: Bánesz L, Kozlowski JK, Eds. Actes du XIIe Congrès International des Sciences Préhistoriques et rotohistoriques. Bratislava. 1991; pp. 181-6.

[12] Broglio A. L'Aurignacien au sud des Alpes. In: Bánesz L, Kozlowski JK, Eds. Actes du XIIe Congrès International des Sciences Préhistoriques et rotohistoriques. Bratislava. 1991; pp. 193-202.

[13] Kozlowski JK. A dynamic view of Aurignacien technology. Trab Arqueol 2002; 24: 21-34.

[14] Conard N, Langguth K, Uerpmann HP. Einmalige Funde aus dem Aurignacien und erste Belege für ein Mittelpaläolithikum im Hohle Fels bei Schelklingen, Alb-Donau-Kreis. Archäol Ausgr Baden-Württemberg 2002; pp. 21-7.

[15] Teschler-Nicola M. Early Modern Humans at the Moravian Gate: The Mladeč Caves and their Remains. New York: Springer 2006; p. 528. [http://dx.doi.org/10.1007/978-3-211-49294-9]

[16] Germonpré M, Hämäläinen R. Fossil Bear Bones in the Belgian Upper Paleolithic: The Possibility of a Proto-Bear-Ceremonialism. Arctic Anthropol 2007; 44(2): 1-30. [http://dx.doi.org/10.1353/arc.2011.0015]

[17] Döhle HJ, Weber T. Höhlenbewohner in Mitteldeutschland? Die Funde aus Rübeland. Kat Dauerausst Landesmus Vor- Frühgesch Halle 2004; 1(1): 197-200.

[18] Blasius W. Bericht über die fortgesetzten Ausgrabungen in den neuen Theilen der Baumannshöhle bei Rübeland im Harz während der Jahre 1898 und 1899. Jb Ver Naturwiss Braunschweig 1901; 12: 83-6.

[19] Blasius W. Die Vogelfaunen in den diluvialen Ablagerungen der Rübeländer Höhlen. J Ornithol 1901; 122: 57-60. [http://dx.doi.org/10.1007/BF02206272]

[20] Diedrich C. Impact of the German Harz Mountain Weichselian ice-shield and valley glacier development onto Palaeolithics and megafauna disappearance. Quat Sci Rev 2013; 82: 167-98. [http://dx.doi.org/10.1016/j.quascirev.2013.08.013]

[21] Diedrich C. Evolution, Horste, Taphonomie und Prädatoren der Rübeländer Höhlenbären, Harz (Norddeutschland). Mitt Verb dt Höhlen- Karstf 2013; 59(1): 4-29.

[22] Steiner U. Zwei neue paläolithische Artefakte aus der Hermannshöhle bei Rübeland /Harz. Ausgr Funde 1964; 9: 17-8.

[23] Steiner U, Steiner W. Zur Genese der Rübeländer Höhlen. Geologie 1969; 18: 198-203.

[24] Völker R. Die Grabungen auf Bärenknochen im Frühjahr 1984 in der Hermannshöhle Rübeland. Ber Mitt Zentr Fachaussch Höhlen- Karstforsch 1985; 85(1): 5-7.

[25] Rosendahl W, Döppes D, Joger U, *et al.* New radiometric datings of different cave bear sites in Germany - results and interpretations. Bull Soc Hist Nat Toulouse 2005; 141(1): 39-46.

[26] Koumouzelis M, Ginter B, Kozlowski JK, *et al.* The Early Upper Palaeolithic in Greece: The Excavations in Klisoura Cave. J Archaeol Sci 2001; 28: 515-39. [http://dx.doi.org/10.1006/jasc.2000.0599]

[27] Nordhoff P. Stable isotope investigations on speleothems from different cave systems in Germany. Unpublished Dissertation, Göttingen, Georg-August-Universität 2005; 149.

[28] Töpfer V. Ein neues Feuersteingerät aus der Altsteinzeit aus der Hermannshöhle bei Rübeland in Harz. Vorgesch Museumsarb Bodendenkmal 1956; 3: 9-11.

[29] Grünberg J. New AMS dates for Palaeolithic and Mesolithic camp sites and single finds in Saxony-Anhalt and Thuringia (Germany). Proc Prehist Soc 2006; 72: 95-112. [http://dx.doi.org/10.1017/S0079497X00000797]

[30] Wojtal P. Zooarchaeological studies of the Late Pleistocene sites in Poland. Kraków. Inst system evol anim Polish Acad Sci 2007; p. 189.

[31] Hahn J. Die Geißenklösterle-Höhle im Achtal bei Blaubeuren I. Fundhorizontbildung und Besiedlung im Mittelpaläolithikum und Aurignacien. Forsch Ber Vor- Frühgesch Baden-Württemberg 1988; p. 26.

[32] Zotz L. Die Aurignac-Knochenspitzen aus der Tischoferhöhle bei Kufstein am Inn. Quartär 1964/65; 15/16: 143-53.

[33] Brodar M. Knochenspitzenfundstellen des älteren Jungpaläolithikums in Deutschland und Österreich. Quartär 1968; 19: 219-37.

[34] Liolios D. Reflections on the role of bone tools in the definition of the Early Aurignacien. Trab Arqueol 2002; 24: 37-51.

[35] Albrecht G, Hahn J, Torke WG. Merkmalanalyse von Geschoßspitzen des mittleren Jungpleistozäns in Mittel- und Osteuropa. Archaeol Venat 1972; p. 2.

[36] Straus LG. Upper Paleolithic Hunting Tactics and Weapons in Western Europe. Archaeol Pap Am Anthropol Assoc 2008; 4(1): 83-93. [http://dx.doi.org/10.1525/ap3a.1993.4.1.83]

[37] Müller-Beck H. Eine "Wurzel-Industrie" des Vogelherd-Aurignaciens. Fundb Schwaben 1965; 17: 43-51.

[38] Feldmann L. Das Quartär zwischen Harz und Allertal. Clausthaler Geowissensch 2002; 1: 1-149.

[39] Diedrich C. Late Pleistocene *Crocuta crocuta spelaea* (Goldfuss, 1823) clans as prezewalski horse hunters and woolly rhinoceros scavengers at the open air commuting den and contemporary Neanderthal camp site Westeregeln (central Germany). J Archaeol Sci 2012; 39(6): 1749-67. [http://dx.doi.org/10.1016/j.jas.2012.01.013]

[40] Diedrich C, Weber T. A Middle Palaeolithic Neanderthal camp and contemporaneous hyena open air den site on a gypsum karst hill at Westeregeln (north-central Germany) within a mammoth steppe environment. Z mitteldt Vorgesch (acccepted).

[41] Delporte H. Les niveaux Aurignaciens de la Rochette. Les Eyzies. Bull Soci d'Etude Rech Préhist 1967; 13: 1-24.

[42] Guillaume-Bordes J. News from the West: a reevaluation of the classical Aurignacien sequence of the Périgord. Trabal Arqueol 2002; 24: 147-72.

[43] Bon F. A brief overview of Aurignacien cultures in the context of the industries of the transition from the Middle to the Upper Paleolithic. Trabal Arqueol 2002; 24: 133-44.

[44] Kloos JH, Müller M. Die Hermannshöhle bei Rübeland. Weimar: K. Schwier-Verlag 1889; p. 76.

[45] Küchelmann HC. Bißspuren von Säugetieren an rezenten und subfossilen Knochen Unpublished Diploma-thesis. Oldenburg: Carl von Ossietzky University 1997; p. 127.

[46] Lyman RL. Vertebrate taphonomy. Cambridge University Press 1994; p. 524. [http://dx.doi.org/10.1017/CBO9781139878302]

[47] Jones PR. Experimental butchery with modern stone tools and its relevance for Palaeolithic archaeology. World Archaeol 1980; 12(2): 153-65. [http://dx.doi.org/10.1080/00438243.1980.9979789]

[48] Greefield HJ. Slicing cut marks on animal bones: diagnostics for identifying stone tool type and raw material. J Field Archaeol 2006; 31: 147-63. [http://dx.doi.org/10.1179/009346906791071972]

[49] Meller H, Ed. Menschenwechsel – Jungpaläolithikum und Mesolithikum. Begleith Dauerausst Landesmusd Vor- Frühgesch Halle 2005; 2: 1-92.

[50] Pohl G. Die jungpaläolithische Siedlung Breitenbach, Kr. Zeitz und ihre bisherige Beurteilung. Jschr mitteldt Vorgesch 1958; 41/42: 178-190.

[51] Auguste P. La chasse à l'ours au Paléolithique moyen:mythes, réalités et état de la question. BAR Int Series 2003; 1105: 135-42.

[52] Vercoutère C, San Juan-Foucher C, Foucher P. Human modifications on cave bear bones from the Gargas cave (Hautes-Pyrénées, France). Scient Ann School Geol Arist Univ Thessaloniki (AUTH). Spec Vol 2006; 98: 257-61.

[53] Münzel SC, Stiller M, Hofreiter M, Mittnik A, Conard NJ, Bocherens H. Pleistocene bears in the Swabian Jura (Germany): Genetic replacement, ecological displacement, extinctions and survival. Quat Int 2011; 245: 225-37. [http://dx.doi.org/10.1016/j.quaint.2011.03.060]

[54] Hörmann K. Die Petershöhle bei Velden Mittelfranken, eine Altpaläolithische Station. Abh Naturhist Ges Nürnberg 1933; 24(2): 1-90.

[55] Clottes J. Chauvet Cave The Art of Earliest Times, Salt Lake City, University of Utah Press 2003.

[56] Morel P, Garcia MA. La chasse à l'ours dans l'art paléolithique. Etud Rech Archéol l'Univ Liège 2002; 100: 219-27.

[57] Hahn J. Die Stellung der männlichen Statuette aus dem Hohlenstein-Stadel in der jungpaläolithischen Kunst. Germania 1970; 48: 1-12.

[58] Trinkaus E, Svoboda J. Early Modern Human Evolution in Central Europe: the People of Dolní Věstonice and Pavlov 2006.

[59] Armand D. Abri Castanet (Dordogne, France): An Aurignacien Site with Bear Procurement. Sciet Ann School Geol Arist Univ Thessaloniki Spec Vol 2006; 98: 263-8.

[60] Mellars PA. The character of theMiddle–Upper Paleolithic transition in south west France. In: Renfrew C, Ed. The explanation of culture change: models in prehistory. London: Duckworth 1973; pp. 255-76.

[61] Grayson DK, Delpech F. Specialized Early Upper Palaeolithic hunters in south-western France? J Archaeol Sci 2002; 29: 1439-49. [http://dx.doi.org/10.1006/jasc.2002.0806]

[62] Grayson DK, Delpech F. Ungulates and the Middle-to-Upper Palaeolithic transition at grotte XVI (Dordogne, France). J Archaeol Sci 2003; 30(12): 1622-48. [http://dx.doi.org/10.1016/S0305-4403(03)00064-5]

[63] Weinstock J. Late Pleistocene reindeer populations in Middle and Western Europe. Bio Archaeologica

2000; 3: 1-307.

[64] Diedrich C. Palaeopopulations of Late Pleistocene top predators in Europe: Ice Age spotted hyenas and steppe lions in battle and competition about prey. Paleont J 2014; pp. 1-34.

[65] Diedrich C. The largest European lion *Panthera leo spelaea* (Goldfuss) population from the Zoolithen Cave, Germany – specialized cave bear predators of Europe. Hist Biol 2011; 23(2-3): 271-311. [http://dx.doi.org/10.1080/08912963.2010.546529]

[66] Diedrich C. Ice Age geomorphological Ahorn Valley and Ailsbach River terrace evolution– and its importance for the cave use possibilities by cave bears, top predators (hyenas, wolves and lions) and humans (Late Magdalénians) in the Frankonia Karst – case studies in the Sophie's Cave near Kirchahorn, Bavaria. Eiszeit Gegenw. Quat Sci J 2013; 62(2): 162-74.

[67] Diedrich C. Ichnological and ethological studies in one of Europe's famous bear den in the Urşilor Cave (Carpathians, Romania). Ichnos 2011; 18(1): 9-26. [http://dx.doi.org/10.1080/10420940.2011.552578]

[68] Münzel SC, Langguth K, Conrad N, Uerpmann HP. Höhlenbärenjagd auf der Schwäbischen Alb vor 30.00 Jahren. Archäol Korrespondenzbl 2001; 31(3): 317-28.

[69] Münzel SC, Conrad N. Cave bear hunting in the Hohle Fals, a cave site in the Ach Valley, Swabian Jura. Rev Paleobiol 2004; 23: 877-85.

SUBJECT INDEX

A

B

Cajus G. Diedrich

D

E

I

K

L

R

S

T

U

V

W

Z

Made in the USA
Las Vegas, NV
03 November 2026

57424115R00100